Elements of a Successful Website

MATTHEW EDGAR

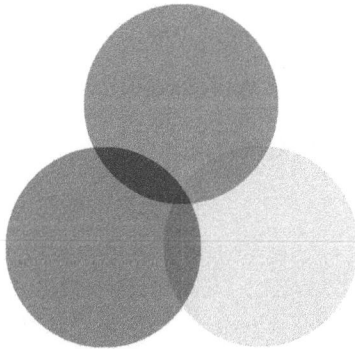

ISBN-10: 0692828060

ISBN-13: 978-0-692-82806-9

Library of Congress Control Number: 2017900730

Matthew Edgar, Littleton, Colorado

www.MatthewEdgar.net

CONTENTS

FIGURES AND TABLES

Figures

Tables

INTRODUCTION

W E ALL WANT THE websites we build and manage to succeed. Regardless of the budget or time we have available, we want as many people as possible to find, visit, and use our website. Most of all, we want the people who visit to move from being a passive visitor into a lead, customer, subscriber, or something of similar value to our organization. This movement from passive visitor to something more meaningful is a conversion—the more conversions, the more success we have.

How do you increase conversions? To start answering that question, let's look at websites that don't get conversions. Over the years of working with clients, I've seen plenty of websites that perform strongly but also many that don't. Of the websites struggling to get conversions, most only change their website once or twice a year (if even that). Your competitors, your industry, your customers' needs, and technology are always changing. Your website must keep pace with all of these changes in order to stay viable.

It's a certain kind of change, though, that leads to success. Failure can happen due to too many changes that worsen the website. The changes that lead to failure seemed like a good idea until they destroyed conversions and potentially damaged the organization behind the website too. The wrong changes made at a large scale are the ones that hurt the most. These large-scale changes waste so many resources that recovery isn't always an option—there just isn't money or time left to fix what went wrong.

In contrast, successful websites are a living, breathing, ever-changing creature. These websites are continually in motion, constantly transforming into something different than they were before. Instead of making large-scale, expensive, and riskier changes, the organizations behind successful websites tend to prefer, and benefit from, steady and slow adjustments.

The changes are measured in order to keep what works and get rid of what doesn't. This doesn't mean every change on the website succeeds. In actuality, it means a lot of changes fail. It's just that these failures are kept small and contained so that one change doesn't bring down the whole system. It's this process of slowly and steadily changing a website, allowing for small failures that can be learned from, that increases the chance of success.

So, you need to make changes. But, what do you change? Broadly speaking, you can focus on improving three main areas of a website:

1. People visiting your website want something from your organization and your website. They have expectations about what your website will deliver and how it will be delivered. By making changes, you can help people get what they want in the way they want it. The more people get what they want how they want, the more they will be inclined to visit and use your website.

2. Your organization has an economic motivation behind the website. There may be a product to sell, leads to generate, information to share, people to get into your store, or donations to collect. Whatever the economic motivation is, by making changes you can learn what it will take to get more people to convert by taking these economically beneficial actions.

3. There is also a technical reality of how your website was built and the funds available to make improvements. While you can improve the technology, you also want to find ways to improve how your website works regardless of monetary or technical limitations.

When walking clients through these areas to improve, the immediate follow-up question I always get—and always struggle to answer—is which matters most. Websites that focus solely on the needs of their visitors while paying little attention to the economic needs of their organization will soon find themselves unable to pay for the website (and, in many cases, pay for their organization as well). Of course, websites focused solely on the needs of their organization while paying little attention to the needs of their visitors will soon find themselves with nobody visiting. If outdated or complicated technology prevents people from using your website or from converting, and you can't find a way to work around these technical problems, failure will soon be on the way.

Each area matters most, and you have to improve all three. Successful websites find a way to strike a balance, improving these three areas simultaneously. The winning balance is one where you can meet the needs of your visitors and the needs of your organization while making your technology perform the best it possibly can.

Frustratingly, one website's successful balance is often the failing balance for another. When I first began helping clients build and improve their websites, I assumed you could copy what one successful website did for guaranteed results on another. After working with hundreds of clients on their websites and studying many more websites, what I've learned is each website is unique.

Even if both websites are in the same industry or serve a similar audience, there will be differences in how that audience is reached, along with differences in the products, services, or information offered. As well, people will visit your website for different reasons and with different expectations than any other website. All of these differences mean you can't just copy some other website and hope for a successful outcome.

You can, and should, learn from what others have done on their websites and use those lessons learned to experiment with changes on your website. There is no shortage of gurus and experts who claim to have a winning strategy—learn from those formulas and try what they recommend. Recommendations from many of those experts and gurus are discussed in this book along with the lessons I've personally learned working with my clients.

But as you test these ideas and as you read this book, remember your results will vary. There is no winning formula guaranteed to work. Instead, this book is about the questions you need to ask, the ideas to consider, examples of what has worked for other websites, and the things to measure so you can find the right balance that will lead to your website's success.

• • •

Conversions and Engagements

Focusing on improving your organization's economic interests can be broadly defined as conversion optimization. A focus on conversion optimization tends to mean you put your organization first and are intent on increasing the number of people who convert during their visit to your website. This often takes the form of calls to action—like pop-ups or ads—encouraging people to sign up for a newsletter, contact your sales team, make a purchase, or donate to a cause.

The problem is some of the people visiting your website may not want to convert—at least not yet. For these visitors, those calls to actions encouraging them to take some type of action will, at best, get in their way, making for an unpleasant visit. Worse, by focusing exclusively on the needs of your organization and constantly pushing for a conversion, your website can become off-putting to nonconverting visitors.

This is why part of making changes and finding the right balance needs to be about satisfying the expectations of the people visiting your website along with satisfying your organization's economic expectations. You need to find other ways for people to use—or engage—with your website beyond just converting. An engagement might be people reading blog posts, watching a video, downloading a key resource, reviewing case studies, scrolling through longer pages, or actively spending time with your website.

Finding the winning balance increases not just the amount of people who convert, but also the amount of people who engage with your website. Getting more people to engage requires more than pushing for a conversion with potentially off-putting call to actions. Instead, this requires actively working to improve your website's usability. Improving usability is a process of making a thing work better for the people who have to use that thing. The thing can be any physical or virtual object—a website, an app, a book, a door, a chair, and so on. In the case of a website, the more it meets the needs of the people who visit, the more those people will be satisfied with their visit to your website and the more satisfied they are, the more people will engage.

Focusing on improving a website's usability and increasing the number of people who engage instead of focusing solely on increasing conversions can seem counterintuitive. As I've explained the need to focus on engagements and usability to my clients, I've been told several times, "I need conversions, not usability." Forget balancing the needs of your organization with the needs of your visitors. Forget the people who aren't ready to convert. Instead, "How do I get more of my visitors converting right now?"

The reason so many websites that focus on both engagement and conversion succeed is because conversion optimization and improving usability aren't opposite focal points—the absence of either one will lead to failure. By improving a website's usability, the website works better and is more satisfying for the people who visit. This gets more of the people who visit engaging with what your website has to offer.

As they engage, people learn about your organization and the products or services provided. The more people learn, the more likely it is that some of those visitors will become interested enough to eventually convert. By continually making changes to your website, you'll find the winning balance that makes your website usable enough that people want to engage and compelling enough that people want to convert.

• • •

This Is for Any Organization

Some of the clients I've worked with (mistakenly) think only large-scale changes lead to success. Because the larger-scale changes are expensive, only big companies have the budget or resources to make them happen. Plus, only big companies have enough people visiting to make the investment worthwhile. These thoughts are a part of what keep people from making changes, leading to stagnate websites and usually failure.

It's easy to understand why people think this way. Many of the examples and case studies for usability and conversion optimization revolve around larger websites with thousands or millions of monthly visitors. It isn't always clear how it applies to a website where the number of visitors and the budget for changes are considerably smaller. As well, some of the best information is scattered across different books, blogs, articles, videos, and speeches, making it difficult to locate all the needed information.

Added to this, most people working on websites at smaller organizations are the owners of the business, sales reps, marketers, or IT staff. Even if they understand the need and value of regularly changing a website, the website is rarely their main focal point, and there is a lack of expertise in knowing what changes should be made.

No matter how many people visit your website, no matter how much money or time you have, and no matter your level of expertise, if you are responsible for improving your organization's website, my goal with this book is to give you a way to strike the right balance with changes to your website so that you can get people to engage and convert within the limits of your existing technology.

I've spent most of my time over the last sixteen years working with smaller companies and nonprofits, allowing me to see usability and conversion optimization from a different angle. This has given me a chance to see how regular and frequent changes can help to improve and grow websites, even on a small budget or a website with only a few hundred visitors a month. After all, a website—big or small—is just a thing people use, and by making it work better for the people who use it, the more the people will be inclined to engage and convert during a visit. The question, to me, is "Given even the most limited of resources, how do I start making changes that will help me strike the right balance to succeed?"

●　●　●

Start Small, Start Simple

In a few minutes per day, you can change one page on your website, or even part of only one page. The process starts by measuring the result of previous changes to learn what has worked (or hasn't) and why the results were the way they were. Those results lead you toward a few changes to make the page slightly better than before. As changes are made, review the results to see which changes got more people to engage or convert. Then, repeat the process.

The biggest worry is the changes you make won't be perfect. But no website is perfect. Every website has bugs, issues, and things that aren't quite right. Think of websites as a spectrum. On one extreme, you have the perfect website where every visitor is satisfied, engages deeply, and converts frequently. This is utopian fiction. No website appeals to every visitor, nor can a website convince every visitor to engage or convert.

On the other extreme, you have a website with no conversions or engagements. Instead of engaging or converting, visitors leave frustrated, angry, and confused never to return. Some of these visitors might even tell friends and family about how awful the website is, hurting the website's future. Sadly, this extreme is all too real and affects far too many websites.

Your website will never be the utopian ideal, but by making changes you can ensure it doesn't exist on the other extreme. The point of making changes isn't about obtaining perfection. Instead, changes let you slowly and steadily move toward making your website more usable and more appealing so that it can benefit your visitors and organization.

• • •

Elements of a Successful Website

There are many different elements making up a website—text, design, calls to action, images, buttons, forms, navigation, and more. It's easy to get overwhelmed thinking about all of the many ways each element could be changed and all the problems associated with each element that need to be fixed. Overwhelmed, procrastination sets in and no changes are made. How do you keep it simple? Where do you start?

To keep the changes manageable, I've found it's helpful to group all of the elements you could change into five broad themes. These themes are frequently discussed in usability and conversion optimization research. The themes are:

1. Keep your website simple and efficient to use.

2. Let people control their visit, but offer appropriate guidance.

3. Maintain consistency, but know when to break the rules.

4. Prevent the errors you can, handle the errors you can't.

5. Remember, your visitors are real people with real needs.

These themes don't offer specific rules for what changes to make. There are no right or wrong answers. Rather, each theme is a general principle with its own set of questions and ideas that will guide you as you make changes and find areas to improve. The questions and concepts within each theme have allowed me to help my clients find the right way to change their website to successfully balance visitor needs, organizational needs, and technical considerations, resulting in increased engagement and conversion rates—success.

Each theme is interrelated, but by thinking of each separately, you can narrow your focus and keep the changes and problems you are attempting to solve from becoming overwhelming. To help think of each separately, each chapter of this book discusses one theme at a time, exploring how it applies to your visitors, your organization, and your technology. My suggestion is to read one chapter, answer the questions asked, and based on your answers, select a few simple changes to make your website slightly better—not perfect, just better than it is now.

After making changes, follow the guidance provided to measure the results to see which changes increased conversions or led to more visitors engaging. The more changes you make, the more you learn what works (or doesn't). Based on what you've learned, find another change to make, then another and another. Keep repeating this process—asking more questions, making more changes, and steadily building on what works while removing what doesn't. The more changes you make the more successful your website and your organization will be.

CHAPTER ONE

SIMPLE AND EFFICIENT TO USE

I**T'S NICE TO THINK** people will be excited to visit your website, and during their visit they will eagerly read and interact with every page and feature you've added. They will become so interested in what you have to offer that they will readily convert. Any troubles or difficulties they have during a visit to your website will be quickly forgiven because of how much they admire you.

In reality, people come to your website with a specific task in mind—placing an order, answering a question, watching a video, contacting your company, getting an address, finding an event to attend, learning something new, or more. Whatever a person's ultimate task is, he or she wants to accomplish it as quickly and simply as possible so that he or she can get on with the day.

At best, your website is a helpful tool allowing people to accomplish their desired task. At worst, your website is a tedious chore a person must suffer through to check that task off his or her to-do list. People are rarely interested in spending their precious time trying to understand how your website works. They don't want to click through a large number of pages or scroll extensively to find what they are seeking. The more trouble people have during their visit, the more likely it is they are going to leave in hopes of finding some other website that will help them complete their task.

These tasks people wish to complete are the various ways people can engage or convert during a visit to your website. As a result, helping people complete their desired task is the same thing as helping people to engage and convert. As you make changes to your website, you want to increase how much your text, design, and functionality helps your visitors. Part of that is finding ways to make the process of completing those tasks simpler and more efficient— such as removing steps from the process or reducing complexity in how the steps are presented.

Key Concepts and Questions

• • •

Simplicity and Minimalism

Simplicity is often compared to and confused with minimalism. While the two are similar, there are differences between these concepts when it comes to making changes to your website. Minimalism refers to a visual style within your website's design, the images used, or the tone taken in your website's text.[1] Simplicity has less to do with design or tonality and more to do with how easy it is for people to interact with your website's organizational structure. To best serve and help your visitors complete their desired tasks, you want to strive for simplicity, but that doesn't mean your website needs to adopt a minimalist style.

More practically, the biggest difference between the two concepts is that minimalism's aim is a reduction in the quantity of stuff included on each page, where simplicity's aim is a reduction in the complexity in the way all that stuff is presented.[2] Your website might have a lot of text, images, features, or functionality contained on each page. All of these items together can cause complexity, making your website challenging for people to use.

One route to reducing complexity is to remove some of the things included on each page—which is adopting a minimalist style. But another route to reducing complexity is to work toward simplicity in how those items are presented. Even a very full website can be simple to use provided it's presented in an orderly fashion. There is nothing wrong with adopting a minimalistic style and reducing the amount of stuff contained on your website. But that visual style may not be appropriate for your organization.

As well, sometimes it isn't possible to remove items from your website. You may have contractual obligations to show certain images, like logos of sponsors. You may also be legally required to include certain passages of text, like disclaimers or privacy policies. There may also be a long list of things your visitors expect to find during their visit—like features they want to interact with, text they expect to read, or images they want to view. Removing items people expect to see when visiting comes with the risk of irritating those visitors if they can't find what they were looking for.

To begin simplifying requires focusing on improving how your website is organized. Take your website's navigation as an example. The links in the navigation may be jumbled together, and the people visiting may not be sure how to find the pages they are seeking. There might be an order to how the links are included, but whatever the order may be, it gets lost in the sheer volume of links. This causes visitors to feel overwhelmed and confused. There could also be redundant and duplicated links, leading to even more confusion.

Many visitors who see such a complicated navigation would probably leave your website instead of spending their time and energy figuring out how to use it. Obviously, navigation is just one of many examples where complexity can arise—every page, every paragraph, ever feature, and every call to action all need to be organized and presented in the simplest way possible. Any disorder or confusion in the organizational structure will inevitably lead to a reduction in conversions and engagements.

The reason so many websites suffer from complexity is bloat. There is a strong urge to add more pages, blog posts, videos, images, graphs, charts, comments, and other features or functionality. All of these items make your website more worthy of visiting, which helps encourage people to visit. The problem is how to present all of these items. Many websites try to include every single page they've added, reasoning visitors need to be able to locate each page somehow so it might as well be promoted. The home page, navigation, sidebars, and footers become crammed with dozens of links with references to every single page, which leads to visitors feeling overwhelmed by the complexity. If every page on your website is treated with equal importance, nothing will seem very important.

Bloat is often thought of on a website as a whole, but it's easy for any given page to become overly complex with too much text, too many images, too many videos, or too many calls to action. Too much bloat makes the page busier, leading to visitors having a difficult time engaging or converting during a visit to that page. Visitors may want to engage or convert, but how would they know which of the five different, equally compelling calls to action to click? People may be interested in what you have to say, but if the page is so visually distracting, it may be near impossible for visitors to read the text you've written.

Visitors will instead favor pages that appear less daunting. That doesn't mean your pages need to be incredibly short or empty. Again, simplification is a reduction in complexity, which is not the same as reducing the quantity of stuff. If one page happens to contain a lot of text, you can simplify this longer page by breaking it into smaller chunks. By using more headers, subheaders, lines, images, pagination, or other design features to break apart long blocks of text, you create smaller and simpler chunks of text.[3] People can view each smaller chunk as a discrete object versus seeing the entire page as one long, complex, and overwhelming thing they must read through.

Complexity is also likely to occur with forms. Whether it's a contact form, sign-up form, order form, or a form allowing some other means of converting, it's possible the form includes too many fields. Too many fields will make the form appear difficult or tedious to use, which leads people to skip filling it out, reducing conversions.[4]

How many fields are too many requires looking at a visitor's perception of complexity. The reality may be the form is easy enough to complete but it doesn't look it. Through design changes, you can simplify the visual complexity of the form and allow more people to feel comfortable using it. For instance, one common design issue that causes complexity is how spread out the fields are. The more spread out the fields, the longer the form looks, and people assume it will take too long to complete it. In these cases, changing the design to reduce the space between fields might help the form not seem so long. In other cases, the fields are so tightly packed together that people can't read the text around the fields. This too makes the form look complex and difficult to use, so people will avoid converting by using that form.

Another solution to real or perceived complexity is to reduce the number of fields. In the case of some forms, a few fields are included strictly for the benefit of the organization. On your website, you might include a sales-inquiry form for people to indicate they are interested in speaking to a sales rep about making a purchase. It might help your sales process if people provide details about their budget or project schedule. So in order to help your sales reps, you include a field on your form asking for each bit of information.

However, adding these fields causes two problems. First, there are now two more fields on the form, and the more fields you have, the more complex things look for the visitor. If nothing else, this makes the field look longer and more time-consuming to complete. But, along with the design issues causing complexity, these new fields might add to the actual complexity of the form. People might not know their budget or schedule and they will struggle to respond to these fields you've added.

By removing the fields people don't know how to answer, you decrease the complexity and increase the chances more people will complete the form. Of course, in this example, your organization may not want people who don't know how to answer questions about their budget or schedule to submit the inquiry form and talk to a sales rep. Without this information, maybe people won't make a high-quality lead for your organization. Also, it may not be worth the person's time to talk to your sales rep until they can answer questions about their schedule or budget. Given that, the appropriate answer might be to accept a lower quantity of converters due to a complex form in exchange for a higher quality of converters.

If, however, your conversions are currently low both in terms of quality and quantity, review the fields included on your website's forms. For each field, ask yourself how much that field increases complexity, real or perceived. If the complexity is due strictly to a design flaw, like spacing issues, then make adjustments to the design. But if the complexity can't be resolved by adjusting the design, then you need to decide if the complexity caused by the field is worth it given the benefits to either your organization or your visitors.

•   •   •

Efficiency

Efficiency is about how much time people spend and the number of steps people must go through in order to complete a desired task. Your visitors typically want to spend very little time interacting with your website. Each step you make visitors take to complete a task slows them down. Anything irrelevant or superfluous will mean your website is inefficient and a waste of their time. So the fewer steps you make people take, the quicker they can complete their desired task. This will leave them more satisfied with their visit. But, it will also increase the chances people will successfully engage or convert—after all, those tasks people want to complete are the various ways people can engage and convert during their visit to your website.

Like with simplicity, you need to consider more than just the actual amount of time or steps it takes to complete a task. Instead, you need to consider the steps and time as perceived by a visitor. An increased number of steps might actually make for a seemingly faster visit. For instance, a form on your website might contain five steps with three questions asked per step. Each step seems relatively quick and easy to complete, so visitors can answer the questions and complete the form without issue—engaging or converting as they do.

In reality, though, it might be quicker to complete all the questions on a single step because that way visitors don't need to click a button to advance to the next step. This difference between perception and reality might be because the singular step containing all the fields looks like it is too difficult to complete, so many visitors simply choose to ignore this form altogether.

Along with improving perceived time, breaking a form into multiple steps can also save the actual time required. Say a visitor is completing a sign-up form and you've configured this form to work in multiple steps. This gives you the chance to modify what fields are shown in later steps based on a visitor's answers to fields contained in earlier steps. Part of these modifications to later steps might include the removal of fields you don't need answered. For instance, because the visitor said they were signing up for a personal account, you can skip the step later in the process asking for information about their company. While this multistep process might add buttons for people to click to move between steps, it actually can make the overall process seem to go by quicker for the visitor.

Of course, making a visitor click to too many different steps can instead increase the perception of how long a task takes. When visitors are placing an order, seeing that there are still five more steps to go can make the order process seem like more work than if they could see the entire process in just one step.[5] In this case, it might be more efficient for visitors if they can see the entire order process in one single step—and visitors may be more likely to convert.

There is no right or wrong way to present a long form on your website. Breaking a form into multiple steps can be advantageous, making your website seem more efficient—or it can be a burden to your visitors, slowing them down. With your forms, try different ways of presenting steps and try different amounts of steps to see what helps the most people complete the form successfully.

A common example when discussing efficiency is the one-click buying offered by Amazon. It certainly appears efficient as this one-step order process reduces the actual time required to place an order as well as the perceived time. Of course, Amazon can achieve one-click buying only because they've long since made customers go through the other steps of the process—gathering billing, shipping, and payment information. That means that Amazon's one-click buying isn't at all a one-step process. Rather it's an intelligently designed multistep process that spreads out the various required steps in such a way that it reduces the perception of how many steps are actually required to place an order with Amazon.

On your website, a one-click buying process may not be an option as you might have many one-time visitors who will not create an account or return to order again. As a result, your visitors might need to take several steps to complete a single transaction—selecting the product and adding it to their cart before inputting billing, shipping, and payment information. But there are still opportunities to simplify each step of the order process and reduce the total amount of steps. For instance, you can ask for the customer's ZIP code first and use that to autofill city and state instead of making the visitor enter their city, state, and ZIP code. Like with the one-click buying, you aren't eliminating these steps (after all, you still need to know the customer's city and state). Instead, you are altering how this information is entered, leading to a more efficient order process.

While the prior examples are for e-commerce website, efficiency matters for any website. If a visitor's desired task is locating your phone number or hours of operation, how many pages will he or she need to click through to find this information? How much time, actual or perceived, is required to click through those pages? If your phone number and hours of operation are located on every page of the website in a prominent location, people will be able to more efficiently access this information. If your phone number or hours of operation are only located on your contact page, people would have to find the link to the contact page, click the link to access that page, and then scroll through the contact page to locate the desired information. That's at least three steps required to get your organization's phone number or hours of operation. If people have to enter their ZIP code to find a location near them before getting the hours of operation or phone number, this adds yet another step.

As you consider how to make your website more efficient, you need to consider what information visitors need to reference and where the information needs to be referenced. People might need to look at your FAQs or a pricing table when signing up for your service or filling out a sales-inquiry form. If people are in the middle of the form when they realize they need to look up a piece of information elsewhere on your website, how many steps and how much time is required to access this information? If this information is regularly needed by your visitors, you can make the form simpler and more efficient to complete by repeating key information or providing links to let people access this information more quickly.

•   •   •

Novice versus Expert

How you make your website simpler and more efficient depends, in part, on knowing how familiar visitors are with your website. How well do your visitors know the words you use within your text? How familiar are visitors with the way you have chosen to organize your navigation? How familiar are people with the way you are asking them to convert?

Gauging the level of familiarity visitors have with your website depends on how frequently and recently people visited your website. The more frequently a person visits, the more he or she will become familiar and remember how your website's navigation, links, calls to action, text, images, and other items are organized and presented. The more recent the visit, the clearer the person's memory will be.

People who are more familiar with how your website is built—an expert visitor—will be able to navigate more quickly even if your website's organizational structure is somewhat complex. Through those frequent and recent visits, they've learned how to bypass at least some of your website's complexity so that they can quickly and efficiently get whatever it is they want. These expert visitors may have grown so accustomed to your website's complexity that making changes to simplify your website may do little to help them. While there still may be ways to reduce the complexity to support expert visitors, these visitors might benefit more from shortcuts you can add to reduce the steps that must be taken—can they place an order without reentering payment information, can they reach advanced support more quickly, or can they access resource materials without being forced to rewatch that introductory video?

On the other hand, people who are unfamiliar with your website—novice visitors—will generally be slower as they navigate and may need to spend more time looking at more pages to understand what it is your organization does and how your website could help them. The text, images, calls to action, and other items need to be designed assuming no prior knowledge of what things do or what anything means. The simpler you can present your text and the clearer you can organize the navigation, the more your website will be able to encourage these novice visitors to stay, engage, and, eventually, convert.

If your website only had to appeal to new visitors, making changes would be easy. After all, novice visitors, by definition, know nothing about your website prior to their arrival. You could make bigger changes to your design or text, and new people wouldn't know what has changed. If the changes didn't work, you could simply change again and the next batch of new visitors wouldn't have any knowledge of these previous changes you'd made. The challenge, though, is that websites have to appeal to both experts and novices. Both groups of visitors have different needs, and yet experts and novices will often visit many of the same pages.

Any changes made to pages accessed by expert visitors come with a high risk. These expert visitors are familiar with those pages already—by definition, their familiarity is what makes them experts. So adjusting links, buttons, design features, calls to action, pages, or something else experienced visitors rely on during a visit may disrupt their familiarity with your website, and they'll have to learn how to use your website all over again. This reduction in complexity can have the ironic twist of making an expert visitor's visit more complex and less efficient. This may make your expert visitors leave your website for good.

This problem of offending your expert visitors can occur during larger-scale website redesigns. In many cases, these redesigns are focused on serving the needs of novice visitors. Maybe novice visitors comprise the majority of the visitors to a website, so naturally the redesign put the needs of these visitors first. As well, during a redesign it's easy to grow excited about the new features you could add that might draw in a completely new audience. Unfortunately, these types of changes can result in large drops in engagement and conversion rates, especially if expert and repeat visitors are more likely to engage or convert than novice visitors.

None of this is to say you should avoid adjusting the parts of your website used by expert visitors or avoid larger-scale changes, like a redesign. But it does mean you should be careful with the adjustments and larger-scale changes. Start by understanding what your expert visitors do during a visit to your website. As you consider changes that will affect these expert visitors, test these changes at a small scale first—instead of reorganizing your entire website, just reorganize one page to see how visitors—expert or novice—respond. Or start by testing those larger-scale changes on a smaller group of visitors. With these smaller tests, you can measure the impact the changes have before rolling the change out to every visitor.

How many expert or novice visitors you have will vary depending on the nature of your website and your organization. If your organization depends on repeat customers for sales, then you may want more people repeatedly and frequently visiting your website. Where the majority of people visiting are experts, the risk faced when making bigger changes is greater. Other organizations may rely on one-time customers and will rarely, if ever, have to worry about how expert visitors might react to a change. In other cases, if there are repeat visitors, there might be considerable time in between each visit, and the people returning will have forgotten what links to click or pages to visit, essentially making them novice visitors.

It can be hard to determine how to develop a new website when you don't know what type of visitor you are likely to attract. Instead of making guesses about what features to develop for a group of possible expert visitors, the better solution is to start a new website by assuming every visitor will be a novice. Initially, this assumption will prove accurate as people visiting a new website won't be familiar with it. As well, assuming visitors are novices is safer than assuming a level of expertise that doesn't exist. Features intended to help expert visitors can confuse nonexpert visitors, making them struggle to use the website. However even the most expert visitor won't have to struggle through using a simply designed website.

After launching a new website, you can monitor what type of visitors you have and what those expert visitors do differently. For instance, you may notice your expert visitors are more likely to use your website's sidebar navigation—suggesting you could adjust the navigation items included in the sidebar to improve how efficiently your expert visitors can access different pages. You might also find only novice visitors look at some of the products you offer on your website, meaning you can test bigger changes with the way those products are displayed.

Technical Considerations

• • •

Device Differences and the Fold

People visit your website on a wide array of devices, including traditional desktop computers (with large, medium, and small screens), laptops (with various sizes of screens), smartphones (yet more screen sizes as well as vertical or horizontal orientations), and tablets (more sizes and orientations). Simplicity means different things depending on the device a visitor is using and the size of the device's screen. A website that looks simple on a larger desktop or laptop screen can appear overwhelming and difficult to use on a smartphone or tablet with a smaller screen.

The solution, though, is not removing everything from your website when it appears on smaller screens. Removing things like images or blocks of text might make your website appear less intimidating or complex on these devices, but it will also take away many of the visual and textual features people expect and need to find. This leads to people on smaller-screen devices spending more time trying to locate the items you have removed, reducing the overall efficiency of their visit. The unfortunate reality of handling your website across multiple devices is that removing text, images, or other items might make the website look simpler, but it doesn't necessarily make using it more efficient.

As you decide how to scale down your website to fit onto smaller screens, you typically want to make everything people want to find during their visit readily available. Some parts of your website will not be as critical to your visitors on smaller-screened devices. For instance, decorative design features can be hidden from view on smaller screens to simplify a visit without decreasing efficiency.

Keep in mind that even these seemingly noncritical decorations can provide visitors with an idea of what they can find on your website and how they can find it. These items might also help people remember your website. The best advice is to remove any part of your website from view cautiously and carefully monitor if the removal has an impact on the amount of time people spend using your website, the number of pages people access, or the number of people leaving your website without converting or engaging.

With whatever text, images, design features, functionality, or other items you decide to keep, they usually need to be structured in a different way on smaller screens. A common approach on smaller screens is stacking items vertically instead of side by side like on wider screens. This can make pages rather long. As the length of your page grows, whether on large or small screens, you are decreasing the efficiency because each scroll visitors must make to reach whatever they are seeking adds another step to their visit.

The good news is research shows people do scroll, but visitors still pay more attention to the top of the page than anywhere else.[6] While 76 percent of people scroll somewhat on pages, only 22 percent of scrolling visitors ever reach the end of the page.[7] Given this, the odds of a part of your website being seen decrease dramatically if it's not located above a visitor's first scroll.[8] This means the things people most need to interact with need to be placed nearer the top of the screen. This helps increase efficiency because people don't have to scroll. However, organizing the pages of your website with the stuff people want to find first on the page also reduces the overall complexity of a person's visit.

To help you limit the amount your visitors must scroll, you want to pay attention to what is included above the fold—or above the scroll—on a visitor's screen. The fold is a term originating in newspapers describing the section of a newspaper you can see without unfolding the front page. On websites, the phrase is used to indicate the amount of the screen people can see without having to scroll. The above-the-scroll or above-the-fold area will change on different devices as larger screens can show more of your website's pages before people have to scroll than smaller screens.

When designing your website, you need to accept some visitors will not scroll to see whatever is located lower on a page. As you decide how to scale your website to fit on smaller screens, you need to decide what text, images, design features, or other items are critical enough that they should be placed nearer the top of the screen.

It's tempting to place calls to action above the scroll to make the process of converting simpler and more efficient—and to ensure visitors don't miss a chance to convert. But, a call to action in the above the scroll area can add complexity. Visitors may have not yet had enough time to understand what your website offers or why they'd want to convert, so a call to action adds complexity by getting in a person's way.[9] This call to action near the top of the page becomes one more thing for people to scroll beyond to find whatever it is they were looking for—assuming people bother with scrolling beyond the call to action at all and don't opt to leave instead.

The better approach is to place the items your visitors want to find and the ones that will most help your visitors understand who you are at the top of the screen. As soon as people arrive on their smartphone or other smaller-screened device, they know what your website offers and how your website can help them complete their desired task. This increases the chances people will stay on your website and engage by scrolling deeper on the page. As people scroll, you can show calls to actions to encourage those engaged visitors to convert. This isn't to say you should never include the call to action in the above-the-scroll area—after all, on some pages the call to action showing people how to convert is what people expect to find.

While you should pay close attention to the area above the scroll, the area below the scroll is not a virtual no-man's land. While not everybody will scroll so low on a page, some visitors will, especially if the above-the-scroll area offers a compelling reason to scroll beyond the top of the screen.[10] Visitors who scroll to this area have engaged with your website and, by engaging in this way, have indicated a desire to learn more about what you do. This makes the area below the scroll an ideal area for placing calls to action—while not everybody will see what you offer here, the people who do scroll here have seen and read more about your organization. The extra information they have learned can make it more likely these engaged visitors are interested and ready to convert.

• • •

Colors

Colors can help to improve the efficiency of your website as they offer a means of communicating with your visitors without having to clutter up the page with more text, images, buttons, and other design features. An offline example of colors improving efficiency is the traffic signal where there is no text telling a driver what to do. Instead, a color scheme acts as the primary indicator of how a driver should proceed at an intersection.

The same can happen on websites. For example, when displaying errors, having the error message in red text or a red background can indicate which errors are critical and require attention immediately. However, yellow text or a yellow background behind the error may indicate the error message displayed is only a warning and does not require immediate action.

Along with these functional uses, colors can also be helpful to attract attention to key pieces of information, like pricing, phone numbers, calls to action, or buttons on forms. When trying to attract attention to particular features, the higher the contrast, or difference, between two colors is better as this high contrast makes these features more prominent and noticeable as people scroll through a page. In fact, you may be able to get more people clicking on a call to action simply by increasing the color contrast.[11]

These contrasting colors can also make it easier to engage. For instance, greater contrast between the background and text colors can help people read your text more easily.[12] This is why black text on white backgrounds has long been a popular choice for reading text. As well, greater contrast can help people with impaired vision read your text more easily. The Web Content Accessibility Guidelines— guidelines for making a website more accessible to all visitors— require a color contrast of at least 4.5:1 for any critical text.[13] For reference, black on white has a contrast of 21:1.[14]

While using different colors can lead to efficiencies and help those with disabilities have an easier time using your website, too many colors can also increase your website's complexity. Websites that use too many colors are overly distracting or simply impossible to read.[15] Further, websites that use too many colors become visually unappealing to most visitors. Instead, limit the amount of colors you use and consider creating higher contrasts by using different tints, shades, or saturation values of a few different colors.[16] This allows you to maintain a consistent, appealing, and satisfying visual aesthetic while still utilizing colors to the benefit of your visitors.

The downside to using color to communicate with a visitor is you cannot rely on it to work for all visitors. Studies have found that 8 percent of adult males and 0.5 percent of females of Northern European ancestry have red-green color blindness.[17] Older visitors without this color blindness may still find it challenging to distinguish between the different colors used.[18] Even if your visitors are younger and have no vision-related health concerns, they might have monitors displaying colors differently or might be viewing your website in dim or bright light where colors and contrasts between colors are not as noticeable.[19] Even if a visitor's eyes, monitor, and lighting are impairment free and the visitor can see exactly the colors and contrasts you intended, he or she may still simply not get whatever it is the color is communicating.

What this means is that while there are benefits to communicating with visitors through color or contrasting colors, color cannot be the only means of communicating a concept.[20] Information needs to also be shared with visitors through the location or shape of the object, as well as the text within the object. Returning to the traffic signal example, the signal communicates with a driver primarily through color, but it also communicates by the position of the different lights within the signal. Other road signs communicate not just with shape, position, and color, but also with the words within the sign.

A common area on websites where you need more than color to communicate with visitors is in links. If the links within your website's text are only identified by a color, some visitors may miss your links, especially if they have some type of visual impairment. To a certain extent, this can be alleviated by ensuring the link's color and the text's color contrast enough where most visitors can identify what is or is not a link. However, given how important links are for conversions and engagements, visitors who do not understand your links can be detrimental to the success of your website. A better solution is to communicate links with more than color, which is why an underline on links is common.[21] By underlining links, you give your visitors two signals to help them locate a link within your text. Even if a visitor never saw or understood the color of the link, the visitor can still click links to engage or convert.

$$\bullet \quad \bullet \quad \bullet$$

Fonts

Part of maintaining visual simplicity and avoiding complexity requires using font styles, font sizes, and font spacing that can be easily read. If you use too many fonts or too many different sizes, your website will begin to appear visually complex and people will perceive it as difficult to use. That isn't to say all of your text should use the same style or size. The aim is to have a design that looks orderly. Using different font sizes and styles can help distinguish headers from the main text or make certain parts of your text stand out. An orderly use of fonts can make your page easier for people to skim through, read, and use.

The next question is what font size to use. There is a certain aesthetic appeal to using smaller font sizes, as it can create a sleek look, but those smaller font sizes might be indecipherable to many visitors.[22] In most browsers, the default font size is 16 pixels, which is the standard recommended size for a website's main text.[23] While this seems large, remember most visitors on desktop or laptop computers will usually be at least twenty inches away from the screen.[24] Tablet and smartphone visitors are usually about a foot from their screen.[25] At these distances, a 16 pixel font size will make your website's text appear to be the equivalent of the size of text in a book held at a comfortable reading distance.[26] At this font size, the line height should be around 1.2 relative units where the font size is 1 relative unit to allow enough breathing room between each line of text.[27]

Another decision point with fonts is whether to use a serif or sans serif font style. Serif fonts include decorative lines and marks to distinguish letters, where a sans serif font removes these decorations. In printed work, the standard is to use serif fonts as research has shown it can improve readability.[28] However, keep in mind serif fonts that are too decorative will be difficult for most people to read. On the web, sans serif fonts are more common and are typically perceived as modern and clean, but some researchers have found large blocks of text in sans serif fonts can actually be more challenging to read.[29]

There are no concrete rules for font style, spacing, or size. As you decide the font style to use on your website, start with standards and accepted conventions and then modify from there. As well, look at your website's text in a variety of styles and sizes to decide what is the easiest to read given the nature of your website's text and the type of visitors you are likely to attract. If your website's engagement or conversion rates are currently low and you are struggling to identify a cause, consider changing your website's font or increasing the font size or spacing as low engagement and conversions can result from people struggling to read your website.

Behavioral Considerations

• • •

Cognitive Load

Cognitive load refers to the total amount of effort used by the short-term memory when completing a task.[30] You generally want to avoid requiring people to apply a great deal of mental effort when they use your website. If your website is difficult to use—with a complex design, complicated navigation, or many steps involved—the more likely it is people will avoid completing an engagement or conversion. To prevent people from giving up, a simpler and more efficient website can reduce the mental effort required. With a more appealing, organized design, and an appropriate number of steps required to complete any given task, your website will become less mentally daunting for people to use.

There are different types of cognitive load to consider. One type is intrinsic, which is the natural amount of difficulty found within completing a certain task.[31] You can do very little to reduce intrinsic cognitive load because some things are just difficult. If a product your organization sells is highly technical, explaining what the product does and the benefits it brings to a customer might be intrinsically difficult, meaning there's little you can do to make it any easier.

However, you can limit intrinsic cognitive load by deciding how to present complicated information. Completing a long form is intrinsically harder than completing a short form because no matter how much you change the design of the long form, there are still a lot of fields people must work through. A cleaner design with orderly labels explaining what each field is, as shown in figure 1-1, can reduce a visitor's cognitive load making even a longer form easy to complete.

Similarly, you can't reduce the effort required to understand a difficult subject matter, but you can decide how to organize the text

Figure 1-1. *Clearly stated labels, like Your Full Name and Phone Number in this example, reduce the mental effort required to understand how to respond to a field.*

discussing that subject matter. By breaking it apart into smaller chunks, you give people a chance to understand little pieces at a time instead of trying to explain the whole of the complexity at once.

This process of deciding how to present intrinsically difficult material to your visitors is the other type of cognitive load, which is called extraneous cognitive load. Extraneous cognitive load is high when information is presented in a complex manner or when superfluous information is included.[32] The simpler you can present the information contained on your website, and the simpler you can make using any features or functionality, the less mental effort your visitors must exert.

The amount of cognitive load required to engage with your website and the complexity people will tolerate will differ depending on the expertise of your visitors. If your visitors are experts at using your website or in the subject matter your website discusses, they will likely be familiar with any inherently difficult tasks.

As an example, a website about tax law designed for people who aren't accountants would need to be written, organized, and presented quite differently than a website about tax law designed for accountants with a decade of experience. In the same way, a website designed for expert visitors completing repeat orders would look considerably different than a website designed for novice visitors completing their first, and only, order.

• • •

Recognition versus Recall

Human brains can quickly recognize things, but at the same time, our brains struggle to recall facts. Recognizing that the phrase "October 29, 1969" is a date doesn't require much effort, but recalling what happened on that date requires more effort. When you see a familiar thing—like a picture of a person you know or a familiar grouping of text—you can retrieve the memories associated with that thing because seeing it triggers similar neural patterns, making this a relatively simple means of retrieving a memory.[33]

Unlike recognition, recalling a fact requires retrieving old memories about something without being able to trigger those similar neural patterns in the brain, which means the brain must work harder to recall that information.[34] In case you are struggling to recall what did happen on October 29, 1969, that was the date the first message was sent across what would grow into the Internet.

To keep your website simpler to use, you want to allow visitors to recognize what they are supposed to do instead of making your visitors recall what they are supposed to do. Let's say your visitors come to your website to shop and as they shop people will probably look at several different products spread out across a multitude of pages. As your shoppers narrow their selections, they may want to revisit some of the pages they have previously seen. Likely, people will not easily recall all the products they've seen, the pages those products were on, or where the links to those pages were located. You can help people recognize their previously viewed products by showing pictures of those products on each page under a header that labels these pictures as "Recently Viewed Products." Each picture could provide a convenient link back to the corresponding product, making the shopping experience more efficient.

A similar technique could be applied outside of an e-commerce website to help people navigate informational websites. If people view your website to access resources and conduct research, your website could remember the different pages they've read and show that list of recently read pages to visitors. This recognition-based technique would allow for an easier means of navigating through previously viewed sections of your website.

For similar reasons, it's helpful to repeat key information instead of assuming people have remembered every detail they read. If your website lets people complete a sales-inquiry form, you probably ask people what types of services are of interest. Unfortunately, the description of those services might be located on a different page than the page containing the sales-inquiry form. To find the description of your services, people either have to recall the list from memory or they have to recall where the link to the list of services was. Either way, this adds a layer of complexity to sending your organization a sales inquiry. If people click back to the page listing the services, this also makes the process of completing the sales-inquiry form less efficient.

Alternatively, people may choose not to click back to the page listing your services and will instead attempt to recall your services, likely with little accuracy. This lack of accuracy would make the submissions you receive from your sales-inquiry form somewhat less desirable for your organization. You can help your visitors (and your organization) by offering a quick summary of your services within the form, possibly along with a link to retrieve more information about those services. By doing so, you would help visitors retrieve this information via recognition instead of recall, which would simplify the process of completing the sales-inquiry form.

What to Measure

• • •

Measuring Mental Effort

To gauge how simple using your website is, you need to measure how much mental effort people exert while engaging or converting. In a lab, psychologists measure mental effort by pupillary response, watching for increased eye dilation, which indicates a task requires higher amounts of cognitive load.[35] This type of lab testing can be very helpful to understand mental effort, but a more practical, budget-friendly metric is needed to track the mental effort your website requires visitors to exert.

One way to measure this is by how long people spent on the website and how many pages were visited. These measurements may say a visitor spent two minutes clicking to three different pages. By itself, knowing visitors spend two minutes looking at three pages tells you nothing about the simplicity of use or the visitor's level of mental effort. This example might be a simple, satisfying visit as somebody was able to skim through just three pages in only two minutes to get what he or she needed. Or, this might describe somebody who had an excruciatingly painful visit, having spent two minutes—that felt much longer than two minutes—hoping to find what he or she was seeking before giving up and visiting another website.

To clarify how simple the visit actually was requires looking more deeply at the pages people reached. If the three pages people visited in the example were longer and would take two to three minutes to read each, then for a visit lasting a total of only two minutes you can probably assume people didn't read those three pages, at least not in full. That may be because the page looked too complex and unappealing, in which case you need to simplify. Alternatively, if those three pages contained lists of products sold, then two minutes to look at all three pages might allow plenty of time to view everything, meaning a two-minute and three-page visit indicates a simple and satisfying visit.

Along with time and pages visited, another metric to help you understand a visitor's mental effort is his or her scrolling activity. As you review scrolling activity, remember that it isn't an immediate indicator of complexity. Scrolling is a natural part of using a website and can be a key way for people to locate the information they are seeking. However, scrolling patterns can also reveal areas where people get lost, potentially because of complexities or inefficiencies.

Scrolling can be measured through analytics tools or via heatmaps (heatmaps will be discussed more in chapter 2). Whichever tool is used, you want to gauge average scroll activity for each page of your website so that you know what parts of the page people are, or aren't, seeing during their visit to your website. An example of a scroll pattern from a heatmap is shown in figure 1-2.

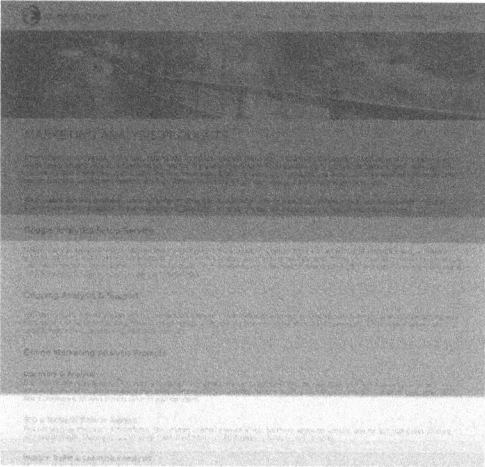

Figure 1-2. A heatmap showing scroll activity. The darker shaded areas indicate more people viewed and scrolled to this part of the page.

A scrolling pattern might show people scrolled through part of a longer page, then stopped a third of the way down before leaving. If most visitors also stop at this part of the page before leaving, it's likely this area of the page is confusing visitors. Perhaps an image located there gives the false sense the page is complete and nothing exists below, in which case maybe that image is adding unnecessary complexity to your design and needs to be adjusted. If you have text in this location, a rewrite to simplify the text in this area may help people scroll deeper. Or maybe the place people stopped scrolling contains a call to action—suggesting either people weren't ready to convert yet or they weren't ready to convert in that way. A different kind of call to action or a different design might work better.

Scrolling patterns may instead show there was an extensive amount of scrolling, but despite scrolling through the whole page, visitors never stopped scrolling long enough to read any of the text, look at images, or watch any of the videos contained on the page. A scrolling pattern like this suggests people were skimming the page looking for something, but nothing on the page caught their attention. It could be that nothing caught their attention simply because nothing on this page was what people were looking to find. However, it could also be there was something on the page people would have like to find, but they weren't able to find it because they couldn't skim through the page. In these cases, you want to find ways to make the page of your website simpler so that they are easier to skim through. By adjusting the headers within the page's text or the images and icons used, you can simplify the skimming process for people so they know when to stop scrolling and start engaging.

Alternatively, visitors to a page may have not scrolled at all but still spend considerable time on a page. This could mean the visitor is confused by the opening of the page and unsure why he or she would continue to read or scroll deeper into the page. More information could be added to the top section clarifying what exists lower on the page, giving people a reason to scroll deeper. For instance, adding a table of contents to the opening of a longer page helps visitors find what they are seeking within that page.

• • •

Measuring Efficiency

Efficiency is measured by how long it takes a person to complete the specific task he or she came to your website to complete.[36] The more time it takes people to complete a task, the less efficient your website will be for visitors, and the less efficient your website, the more people will leave without completing a task. This reduces conversions and engagements because these tasks people are spending time completing are the ways people engage or convert.

Along with efficiency, you also want to measure the effectiveness of the task completion. Where efficiency is concerned with how quickly visitors complete tasks in a certain period of time, effectiveness is concerned with the accuracy and quality of the task's result.[37] While effectiveness is different from efficiency, accuracy and completeness can be helpful measures for efficiency as well. For example, a high amount of inaccuracy in contact form submissions could suggest a problem exists within people's understanding of how to enter data into the fields on that contact form.

The first step to measuring efficiency is to know what tasks people will complete during a visit. Many organizations, big and small, only measure the bigger conversion and engagement tasks—like placing an order, submitting a sales-inquiry form, commenting on a blog post, or subscribing to an e-mail newsletter. For each task, you can measure how many people completed the task and how efficiently and effectively people complete it. Along with bigger tasks, you also want to measure the smaller, micro tasks since not everybody will complete those bigger, macro tasks.[38] Those smaller, micro tasks might include connecting with your organization on social media, clicking on links in your website's navigation, conducting a search, or reading a key page.

As you think through everything people can do on your website, group all the various ways people can engage or convert into macro- and micro tasks. Keep in mind, these groups of tasks shouldn't just include the tasks your organization wants people to complete during a visit. You also want to consider everything your visitors will want to do. By tracking both, you can ensure your website is satisfying the needs of your organization and your visitors. Within your analytics tool, you will want to set up ways to measure how many people completed all of these various macro- and micro tasks, typically in the form of tracking goal completions.

Once you have a list of tasks and a way to track the completion of these tasks, you next want to review what steps people need to take to complete any one of those tasks. For some tasks, especially macro tasks, the steps may be obvious, like in an order process. During an order, people add an item to their cart, go to the cart, input billing information, input shipping information, input payment information, and then are taken to some type of confirmation message. If you have tasks with clearly defined paths, you can measure how much time it took people to move between steps to understand the efficiency of the process. On whatever step people are taking the longest, you can make adjustments to reduce complexity and increase efficiency. As you review the steps required, you can also see what opportunities exist to consolidate the steps or better present the steps.

However, many tasks will lack such neatly defined steps. A micro task might involve a visitor finding the Contact Us page. The steps people take to find this page can vary from visitor to visitor. Some people will arrive directly on your Contact Us page after finding it in a Google search result, allowing those visitors to complete the task very efficiently in a single step. Other visitors might click a link to the Contact Us page from the home page. Still other visitors might miss the link to the Contact Us page that was available on the home page and instead click through fifteen other pages before finally finding the link to the Contact Us page, making that a far less efficient visit.

In many cases, the steps and time required to complete a task should shorten over time as you improve your website. This is especially true where tasks involve accessing information where you want people to get this information as quickly as possible. In the example of reaching the Contact Us page, you would want to change your website to make it easier for people to find links to that page so they can reach it in as few steps as possible.

In other cases, you might find longer times and more steps result in more effective—higher quality—task completions. Maybe a visitor spent a lot of time browsing through your website before placing an order or submitting a sales-inquiry form. On the surface, this suggests an inefficient completion of those tasks. But with all the extra time spent, the visitor learned more about what your organization offers and why he or she should order that product or speak to your sales rep. When the person eventually does complete the task, he or she has a better reason to convert. This makes the person's visit more effective and satisfying to him or her. Likely, this also makes the conversion more beneficial to your organization as well because you now have a customer or lead who is more interested in working with your organization.

Supporting Measurements

• • •

Repeat and New Visitors

Given the different ways novices and experts use, engage, and convert while visiting your website, you need to know how frequently people visit and how recent each person's last visit was. Most web analytics tools provide a way to measure how many of your visitors are new and how many are returning. At a high level, this ratio gives you an idea of who your visitors are and how you can approach updates to your website. You can also determine this ratio for each source leading people to your website—people arriving from Google might attract more repeat or expert visitors while your advertisements on social media attract novices.

Beyond reviewing this ratio of new to returning, it's also helpful to segment your visitors so you can review the differences between each group. This segmentation lets you see what pages visitors familiar with your website look at compared to novice visitors. There will be some pages repeat visitors access but new visitors don't, and vice versa. The changes made to the pages accessed by your repeat visitors can help those people connect more deeply with your organization and could also help those repeat visitors find more ways to convert— like other products or services they could purchase or other resources your organization offers. The changes made to pages visited mainly by novice visitors could focus on drawing people in and showing people what your organization does.

As discussed earlier, the harder pages to adjust are those visited by both new and repeat visitors as a change intended for an expert visitor may confuse novice visitors. As you segment your data to review what novices and experts do, you will often find there are at least a few areas where their needs overlap. As you make changes, you can find ways to better serve these overlapping needs, reducing the risk of benefiting one group at the expense of the other.

There is a word of warning when using this metric. The data about new or returning visitors is based on cookies tracked within the device and browser the person used to visit your website. A repeat visitor is more technically defined as a person who comes back to your website from the same device and the same browser without having cleared cookies. Not everybody will use the same device or browser, and some visitors might clear their cookies. This means the reported number of repeat visitors will skew lower than it actually is—occasionally by a lot, but mostly only by a little. You want to use this data as a rough estimate instead of assuming it's an accurate portrayal of your novice and expert visitors.

* * *

Task by Devices and Screen Sizes

People visit your website from many different devices, and each device can change how effective, efficient, or simple a visit might be. More will be discussed about these different devices in other chapters, but as an immediate example you may find people using a smartphone take more time completing tasks than people using a desktop computer, or may take more steps in order to complete those tasks. Or maybe people using a tablet are less likely to complete any task. By removing steps or making those steps less complex, you could improve the overall efficiency of a visit, allowing visitors to complete more tasks regardless of the device used.

While differences exist between people using different devices, there are also differences in how people use your website depending on the device's screen. Screen resolutions will be discussed more in chapter 2, but a visitor using a laptop with a small screen may have a more complex and likely less efficient experience. For instance, they might need to scroll more to read text than a visitor using a desktop computer with a larger screen. Simple adjustments to the positioning or the size of text or images on smaller screens might reduce the scrolling required.

You want to visit and test your website with the most common devices your visitors use. This lets you walk in the visitors' shoes, trying to engage or convert as they would on a particular device at a particular screen size. As you do this, you will usually find areas where your website is too complex or where too many unnecessary steps are included. You can make changes to reduce the complexity or remove steps from the process on certain devices. This will improve the overall efficiency and simplicity on any device or any size screen visitors choose to use during their visit.

Evaluate Your Website

Efficiency

- What are the micro and macro tasks people expect to complete during their visit to your website? Remember, tasks are the various ways people can engage or convert.

- If people started using your website on the home page, how many pages, clicks, scrolls, and other steps are between there and end of a particular task? Repeat that question for all the pages people use as an entry point to your website.

- Is each step, click, scroll, and page essential to what the visitor wants to accomplish? Could any given step be consolidated with another or removed entirely? If a step is added for the benefit of your organization, such as requiring people submit a particular piece of information they'd rather not share, is the impact on visitors worth it?

Simplicity

- On mobile devices, can text be read and links clicked without horizontal scrolling or zooming?

- On all devices visitors use to access your website, where is the fold? Can the primary information and calls to action be seen without needing to scroll vertically?

- Are key facts repeated where people need them, or do people have to memorize those facts in order to convert?

Effectiveness

- What do people do next after completing a given task? What tasks would you prefer visitors complete next?

- For order forms, contact forms, sales-inquiry forms, or other areas where people submit information to your organization, what is the quality and accuracy of those submissions?

- What changes could you make to the form or the text around the form to improve the accuracy or quality of a person's submission? If these changes reduce the quantity of conversions, is the gain in quality worth it?

Measurement Guide

For guidance setting up the tools for these measurements, see:
http://www.matthewedgar.net/elements/simple

Baseline Setup	Define tasks you want people to complete during their visit. These are the ways people can engage or convert. These should include smaller, micro tasks (such as reading your about page) or bigger, macro tasks (like registering for an event). Determine how long you expect people to take to complete each task (at least approximately).
Monthly or Quarterly	Measure the tasks people complete on your website. How long do people take to complete the tasks, and how many pages or steps are required to complete a task? Are people taking more time than expected—if so, what changes can you make to reduce the time or steps required to complete a task?
	Check how long people spent on each page. Did people spend too little an amount of time to actually read the page or too much? What changes could help people read the page more deeply and spend more time with the pages?
	How frequently and recently do people visit your website? As possible, review the pages novice and expert visitors look at and the tasks each group completes. Determine the differences and similarities between both groups.
Before Major Changes	Review scrolling activity on each page. How far do people scroll before abandoning your website? What parts of a page are causing visitors to stop scrolling and engaging?
	Review completions by device type and the size (or resolution) of the visitors' screen. Do visitors on certain devices take longer to complete a task?

CHAPTER TWO

ALLOW CONTROL AND OFFER GUIDANCE

P EOPLE VISIT YOUR website because they need something. During their visit, they want to get what they want, when they want, how they want without interference from your website.[1] The more control people have to freely move about your website instead of being forced into something, the more people will see how engaging and converting can satisfy their needs.

But your organization wants things from the website too—namely, conversions. It's tempting to think you'd be better off limiting a person's control and instead pushing visitors immediately toward a conversion. The problem is if you offer too little freedom and control and too much coercion, people aren't going to be satisfied. Once people haven't found what they want and see no choices except the one your website is forcing on them, they will leave to get what they want elsewhere.

Of course, allowing control and giving your visitors freedom to engage and convert however they wish can be overdone. If you offer too much freedom to move around your website, people may get lost. There is a difference between "leave me alone to do what I want, how I want" and "you left me too alone; now I'm lost and don't know how to get what I need." Whether frustrated and confused by too much or too little control, people will not be satisfied with their visit.

There is value in letting people take charge of their visit, engaging and converting as they see fit. But allowing people to take control of their visit requires a balanced approach. You need to offer enough control where visitors are able to pick among many different options, including some opportunities to convert. But you also need to offer enough structure so people understand how using your website and converting will help them meet their needs.

Key Concepts and Questions

• • •

Providing Guidance and Structure, While Still Giving Control

One of the main ways you can provide structure and guidance to your visitors is within the navigation. Visitors rely on the navigation to see what pages exist and which of those pages might meet their needs. If the navigation provides links to the most common information visitors are looking for, your navigation becomes a helpful tool that supports your visitors as they decide where they want to go.

If instead the navigation features nothing of interest to your visitors and only shows pages your organization wants people to find, people would see no value in using the navigation. Instead of helping them get what they want, the navigation is forcing them into whatever it is your organization would prefer they do. In these cases, visitors would have to find another way to move through your website—assuming they still see value in using your website at all.

Of course, only showing the pages of interest to your visitors isn't a realistic option. People may not be interested in finding a particular category of products your organization sells, but are you really prepared to remove that entire category from the navigation? Your organization may have a partnership with another organization that needs to be promoted, even if few visitors will be interested. There may also be pages you legally need to present. Instead of only showing the pages visitors want or only showing the pages your organization wants, you need to strike a balance between the two. As well, there will usually be pages both you and your visitors would like to see included in the navigation.

As you decide what to include in the navigation, it's easy to decide that the navigation should include every possible option. Offering every option avoids having to decide what your organization wants or what your visitors want—give visitors every option so they can figure out what to do. This approach turns your website's navigation into a jumbled mess of links that no visitor will understand. Instead of giving the visitor a helpful tool to use your website, your navigation becomes a deterrent.

Instead, the better—and harder—approach is to decide which options to present that are of most value to your visitors or to your organization and that are worthy of inclusion in the navigation. This makes for many tough decisions about what to include and requires a careful balancing act to decide how best to satisfy the needs of your visitors and your organization within the navigation.

You may find people visiting are very interested in clicking to a page letting them search your products but very few visitors are interested in viewing a page listing all available products. However, your organization's vendors would like to see that list of all available products prominently featured. To best support your visitors, you might decide against those vendors and choose to include a link to the page letting people search products in your website's navigation without including a link to the full product list as your vendors might prefer. This decision might increase conversions—in this case purchases of products—since you are giving visitors more control and ability to access something they want. While your vendors may be upset your website doesn't link to the full list of products, the increase in purchases of their products could hopefully appease them.

There may be times, though, when you should not decide in favor of what your visitors want. A law firm's blog might include incredibly popular recipes written by the firm's staff. While initially written as a way to humanize the lawyers and never intended to garner much attention, the recipes took off and now are of great interest to many visitors. Clearly, visitors would prefer the recipes be given prominent focus within the navigation, allowing them greater control in accessing those recipes. But, understandably, the law firm would rather not include those options as it simply isn't in line with their brand or the image they wish to portray.

As you decide what options to include or not include, you may end up with a long list of pages that should be included in the navigation but there simply isn't enough room to include every page. Even if every page is of interest to visitors and its inclusion would result in more conversions and engagements, including that many pages in the navigation would be impractical and would overwhelm visitors.

Another way to offer control over navigating a website with many pages may come in the form of a search feature. By giving people the ability to search through your website, they can find whatever it is they are seeking. If many visitors use the search tool, you can offer even greater control by adding advanced search tools that allow people to filter through your pages to find exactly what they want.[2]

As you build out the search feature, though, you want to consider which pages are returned within search results for the various search terms your visitors are using. Like with the navigation, there may be some pages visitors will be more interested in finding after conducting a search, and there will be pages your organization is more interested in returning within a search result. For instance, when searching for a course your organization provides, visitors may be more interested in finding a course syllabus or reviews from prior students. Your organization is more interested in returning the sales page to let people register for this course. As with the navigation, you need to decide how best to strike the balance on which pages to return for a visitor's search.

Both search tools and navigation will be useless in helping people control how they navigate your website if these tools don't think like your visitors think. After deciding what pages to include in the navigation or what pages should be returned within search results, you need to decide what words and phrases to use to describe these pages. As much as possible, your navigation needs to use the same words and phrases your visitors use to describe the information, products, and services offered.

As an example, you might describe a product you sell by an acronym, but your visitors describe this same product using a slang term. If your navigation only shows the acronym or if your search tool only returns results if people type in the acronym, people won't be able to navigate or search your website. By forcing people to use the acronym, you have effectively punished your visitors for trying to control how they intended to get what they wanted. But if the navigation uses the slang term for the product and the search tool returns results when people enter this term, your website rewards people for controlling how they use the website.

Another area where people need guidance and structure is before clicking links that lead them away from your website.[3] If people thought the link would take them to another page on your website, but instead they arrive on an entirely different website, you have taken control and sent the visitor away from his or her desired course. In other words, people need to know what to expect of links they are clicking. There are technical ways of notifying visitors, like showing an alert or pop-up when a link to another website is clicked or by adding a symbol next to the link. A simpler, nontechnical way of notifying visitors is to alter the text of the link to indicate the link will take them elsewhere.

As an example, if you link to a news website that reviewed your organization's services, you could word the link text in an unguided manner, such as "Read reviews." This link text doesn't clearly tell people that by clicking this link they will leave your website and be taken to the newspaper's website instead. You could provide more structure within this link by saying something like "Read what the local news had to say about our new services." This link text makes it clearer to people that by clicking the link they will be taken somewhere else. This clarity lets visitors have greater control over what to do with the link. Some visitors may not want to leave your website to view the local news website. But other visitors may be more inclined to click the link because they would like to read what an authoritative, third-party had to say about your organization.

. . .

Bail Outs

When something goes wrong or when visitors change their minds, your website can offer more control by offering a bail out.[4] If people start heading down a path they later decide they'd rather not take, bail outs provide a way for them to take control and reverse their actions. A link back to where they were before, a reset button, or an undo option all offer a way to help people reverse an unwanted action.

A common example happens on forms. Let's say visitors need to fill out an involved contact form on your website where there are multiple steps and many fields. It's probable some people will change their minds partway through and will want to stop completing the form. Some people may close the browser to stop the form submission process, but others may worry closing the browser won't work because information could have already been collected. To alleviate visitor worry, you can provide a way for people to cancel the entire form submission. This option keeps people in control of how their interaction with the form ends. This allows even the visitors who terminate the form submission to leave your website satisfied.

While typically thought of as a complete exit from the action being taken, bail outs can also assist in recovery from temporary deviations. People may still want to complete a certain action on your website, but they'd like to go somewhere else first. When filling out that involved contact form in the previous example, people may need to go to some other page of your website to look up information—like pricing or some detail about the service or product they are contacting you about. On many websites, leaving the page the contact form is on will erase the information a visitor has already entered. Erasing a visitor's entries effectively punishes them for taking control and making a choice about where to go on while filling out the form. If people commonly need to reference information elsewhere on your website while completing the form, the better answer might be to build the form to remember information even if people browse away briefly. That way, your website supports people's choices to fill out the contact form however they wish.

Bail outs can also work in undesired ways for visitors. If people click to a PDF file from your website, they may not realize that PDF opened in the same window or tab as your website. Once done with the PDF, instead of hitting the Back button to return to your website, people bail out of the PDF file by closing the tab or window. In doing so, they have lost the ability to hit the Back button to return to your website. The visitors may not return to your website at all, but if they do return, they have to go through more steps to do so, digging through their history to locate the link to your website or the link that originally brought them to your website. In this particular example, if the PDF opened in a new window or tab, you could give people the ability to bail out of the PDF by closing the tab without mistakenly closing your entire website.

This discussion of bail outs and control tends to suggest no action on your website should be irreversible. But some actions on your website *are* irreversible, like placing an order or submitting a form. The key to remember is that just because an action can't be reversed online doesn't mean it can't be reversed offline. After an order is placed, your organization's support staff can probably help people who accidentally placed the order or made a mistake within the order. Even if the bail out can only happen offline, your website should provide information on how to reverse a mistake, such as contacting your organization for assistance.[5]

If there are truly irreversible actions on your website with no hope of undoing the action once it's completed, online or offline, prior notifications need to be provided to visitors before they take that action. A simple confirmation message asking, "Are you sure?" can prevent mistakes from occurring. This tends to only matter when the consequences are high. If a person clicks the "Remove Items From

Cart" button, it's worth confirming if this action should be taken and reminding him or her that there is no undo. After all, the person might have clicked the "Remove Items From Cart" button by mistake and didn't mean to clear the twenty items currently in his or her cart. On the other hand, if people attempt to clear a simple search form, a confirmation isn't necessary since the search can be repeated easily.

●   ●   ●

Striking a Balance: Conversion Optimization

One of the harder aspects in giving people visiting your website control and freedom to move around is balancing a visitor's needs with the economic needs of your organization. Your website exists, at least in part, to get people to take some type of action. What the action is will vary from website to website, but ideally, you want some percentage of visitors (hopefully a large percentage) to buy a product, submit a form, call your sales rep, sign up for a service, donate money, register for your event, or something similar. Getting people to take this kind of action can require a bit of coercion and convincing, which seems to be at odds with giving people freedom and control to move about your website.

On the one extreme, a case can be made that there are times where you need to take away the visitor's control entirely in the name of what your organization needs. The prime example is on landing pages that you send people to from ads, newsletters, or other marketing channels. Neil Patel, a leader in conversion optimization, suggests that on landing pages you should remove all navigation as a way to "take the lead and control [visitor] behavior" because the people visiting are "not supposed to go anywhere else" other than to convert.[6]

Removing other choices eliminates distractions. The only thing people can do from this page is to convert. One such distraction is the navigation, and studies show by removing navigation, you can increase the number of people converting.[7] But removing navigation options can also annoy and frustrate the people visiting your website as they wonder why their only available choice is to convert. The visitors might want to do something else on the website not related to the conversion, or they may want to learn more about your organization before converting. If there is no navigation, people will be unable to find what they wanted.

Because removing navigation or other distractions can drive people away from your website, this can hurt engagements and conversions long term. Some of those people who leave might have been interested in converting in the future, but your distraction-free website kept them from learning enough about your organization to keep them interested. Because they didn't learn enough about your organization during their first visit, the odds they will return to your website in the future are considerably lower.

Another area where organization and visitor needs can conflict is registration when placing an order. Some websites require visitors to sign up for an account prior to submitting the order. With the registration information, the business can easily contact customers, handle support issues, or promote future sales to those customers. Requiring registration is one of the top complaints in e-commerce usability research because many people would rather not create yet another account, especially for websites where they don't intend to shop again.[8] After all, few people visiting your website want to receive future sales e-mails—they get enough already.

A more balanced option is to make the registration optional or offer a brief explanation of the benefits of registration. This would keep visitors in control of deciding how they want to interact with your organization after placing their orders. This might reduce the overall number of people who register for an account, but people who choose to create an account are indicating they are more interested in your organization than people who didn't. Would you rather stay in touch with people who are interested in what you offer or force everybody to stay in touch with you regardless of their interest level? The people who voluntarily register for an account are probably more likely to become long-term customers, repeatedly working with your organization in the future.

None of this means you should never push people toward a conversion on your website. You need conversions to justify your organization's investment in the website. For many organizations, a website is the main way of connecting with customers, clients, donors, or partners. If you make the decision to reduce visitor control and freedom in the interest of conversions, you want to do so fully informed and aware of the amount of control and freedom you are asking your visitors to give up. You also want to be aware of the longer-term risks you could be taking.

But, you can increase conversions without reducing a visitor's control of your website. Nothing about offering people control and freedom prevents conversions. If anything, by supporting more freedom and more control, you reduce frustration because visitors can do more of whatever they wanted to during their visits. This increases the odds visitors will spend more time on your website learning about your products and services. The more time people spend on your website, and the more they learn about what your organization offers, the more likely it is those people will see the value in converting, either now or in the future.

This is where offering guidance becomes critical. You want to help people get what they want to get from your website, but you also want to help people move from the information they were initially seeking to the next step—the conversion. After all, if people are coming to your website to learn about certain types of products or services that you just happen to sell, the odds are good that some of those visitors will be making a purchase.

Instead of forcing people into a purchase, you can help people do their research about the product, answer whatever questions they have, and help them understand how to make that purchase. When these people are ready, you can offer them a choice of taking an action they were already interested in taking. Purchasing the product doesn't need to be the only choice you offer to visitors—it can be one of many choices that visitors can choose to take when they decide they're ready to convert.

As people move closer toward the conversion point, you can start to remove options to help people stay focused on the task they've chosen to complete. After somebody has chosen to make a purchase, you aren't coercing by removing options. Instead, you are helping people stay focused to do what they have freely decided to do. Of course, you also want to support people who change their minds by offering a way back—a bail out.

If your conversion rates are low right now, instead of looking for ways to reduce control and freedom and force people into whatever the conversion is on your website, consider adding more options. If the options are well structured, using the language visitors use, and think like visitors think, these options can help guide people through your website, eventually leading at least some of those people toward a conversion.

Technical Considerations

• • •

Navigation Design: Clicks and Touch

Giving your visitors control and letting them freely move through your website requires a well-organized navigation that supports and guides visitors through everything your website has to offer. The biggest challenge, as discussed, is knowing what to include and exclude. With each new page you add to your website, the question about the navigation becomes harder. Should that new page be added to the navigation? If so, does another page have to be removed to make room? Or does a less important page need to be moved into a sidebar or footer navigation area instead of receiving promotion within the website's main navigation? Some pages, though, are too important to be removed or relegated to a less prominent position, so where do these pages go?

A common approach to this "too many pages" problem is to add pages into drop-down menus. With drop-down menus, only a few links referencing a handful of pages are visible by default. If a visitor moves their mouse over those visible links (or, in some cases, clicks or taps these links), a submenu appears containing more links. The submenu's links are (usually) connected thematically to the top-level link people hovered, clicked, or tapped. Drop-down menus allow you to present a large amount of pages on your website without showing every single link by default. When done correctly, drop-down menus can help people find additional pages of interest, giving them more options and control in navigating your website.

Poorly designed drop-down menus, however, limit a visitor's control. One of the more common problems is visitors aren't always aware submenus exist because the design of the top-level links in the navigation—the links that are visible by default—didn't look like something that could be hovered over or clicked on. If people do expand the submenu, it might be difficult to use the precise mouse or finger movements required to keep the submenu from accidentally closing before a visitor selects a link.[9]

Another problem is many drop-down menus take a kitchen-sink approach and contain too many levels and too many links. Every new page is included somewhere in the navigation. In many cases, there are three, four, or even more levels of submenus containing more and more links. This is almost always too many links. Even if people are able to identify that they can hover over or click on the top-level links, and even if they are able to traverse the many levels of the subnavigation, it's very likely visitors will be too overwhelmed and confused by all the links to efficiently and effectively use it.

When these problems occur, the drop-down menu will limit people's control by keeping them from learning what pages exist and which pages might be of interest to them. When designing drop-down menus, the links in the top-level navigation should be designed in such a way that people know they have the option to find more links by clicking, tapping, or hovering. As well, it's advantageous to design the sublevels of the drop-down menu to be wider so people can move through the links within the subnavigation more easily.[10] Finally, this is also why there should be strict control over what links are worthy of inclusion in the navigation and which links will only distract visitors and should be removed.[11]

Drop-down menus present a different set of technical challenges on touch-based devices with smaller screens, like smartphones or tablets. On these devices, a visitor doesn't have a mouse so he or she can't hover over links. Also, the smaller screens can make it more challenging to read the links contained in the drop-down menu. Many solutions exist, but the more popular solution is to skip the drop-down menu and hide all of the website's navigation behind a hamburger icon, or the little icon with three lines. Unlike drop-down menus that keep some links visible on a top level, these menus hide all links from initial view.

Assuming visitors know to locate the navigation by clicking on the hamburger icon, this would do little to impair a visitor's ability to navigate the website. Unfortunately, not all visitors understand what the hamburger icon is and, even if they do, the icon may be lost amid the rest of the website's design.[12] It's because of these problems that hidden navigation has been shown to reduce the number of people using the navigation by almost half.[13] Without a way to easily locate and use navigation links, visitors can't easily take control of how they navigate and interact with your website.

There are ways to change the design of the hamburger icon to draw more attention to it. For instance, some tests show by putting the icon inside a border or by labeling it with the word Menu—as shown in figure 2-1—you can increase the people using the navigation on a smartphone or tablet.[14] It's likely the current lack of use seen with the hamburger icon will change as people grow more familiar with it. Or a new, alternative convention might be established replacing the hamburger icon entirely. Regardless of what the future holds, the underlining principle to remember as you change your website is to make sure people can easily and effortlessly locate your navigation regardless of their device.

Figure 2-1. *Various design options for using the hamburger icon for navigation menus. Bar image credit: Font Awesome fa-bars icon.*

The other key to remember is that including links in your main navigation is no guarantee people will find the links to various pages. For whatever reason, technical or otherwise, people may simply miss the links or skip over the navigation. As a result, when adjusting your website and deciding how to promote pages, remember the main navigation menu is only one part of your website's navigation. There are other ways people can find all the pages located on your website—like links in a sidebar or footer navigation area that might be more noticeable than the main navigation at the top of the screen.

Another critical part of your website's navigation is the links contained within a page's text. By placing links to pages within the text of another page, you will greatly increase the chances people can find those other pages on your website. If the link included in the text is relevant to what is being discussed and what would be of interest to the visitor, the chances of people using those links increases even more. So if some pages of your website aren't getting visits despite being included in the navigation, try adding links to those pages within other pages of your website. By doing so, you'll expand the different ways people can find those pages and offer your visitors more control in how they navigate through your website.

• • •

Font Sizes, Zooming, and Images

One of the more basic controls your visitors expect is the ability to adjust how your website looks to suit their particular needs.[15] One way to support this technically is by letting visitors change the font size to make your website more readable. After all, some people will prefer increasing the size of the text so they can see the words more easily. One option in support of giving visitors this control is to create your own functionality for increasing or decreasing the size of the font.[16]

Most modern browsers allow people to adjust the size of web pages, so building this functionality for your website is typically no longer needed.[17] Instead, now your website's design needs to support the scaling offered by modern browsers by defining your website's text, images, tables, buttons, links, or other items people may want to resize in relative units.[18] The more popular relative units are the em, rem, and percent. Unlike a fixed unit such as pixels or points, these relative units allow your website to scale and maintain appropriate dimensions if people choose to zoom.

Along with the technical supports, you need to test what your website looks like if somebody were to resize the text or other features. As each browser handles this somewhat differently, test this on different browsers and different types of devices. If people choose to increase the font size, you want them to still be able to use your website, click on every link, utilize the navigation, and read the text. But if increasing the size of the entire website makes certain objects overlap and blocks the text from view, you are essentially punishing people who decided to look at your website in an alternative way.

Changing the size of the website plays an even bigger role on mobile devices where zooming in is a common behavior. Mobile website best practices defined by Google states people should not have to zoom into your website on a smartphone in order to read text or tap on links.[19] This means by default, before zooming, your website needs to be large enough for most of the people visiting your website to comfortably read and tap.

Although by default most people shouldn't have to zoom in, this does not mean you should prevent people from zooming, even though technically there are ways to do so.[20] Depending on who your visitors are, some people may still wish to zoom in to read your text, even if the default is largely readable and easy to use. Other visitors may find it more convenient to zoom in to clearly see the link they are about to click or to review the details of an image. Others might have bad eyesight and want to zoom in to the text for easier reading. While you should design your website to not require zooming for the majority of visitors, disabling zooming altogether takes away a level of control from people who need or want it.

Whether on desktop computers, smartphones, or tablets, one of the bigger problem areas presented by zooming in to a website are images that contain text. These images do not always resize in a way that makes it easy to read the text contained within that image. As the visitor increases the size of the website and the image gets bigger, the text within the image gets blurry and becomes hard to read. In order to use your website, visitors have to zoom out to see the images, but zoom back in to see the rest of your text. They are being punished for making a choice of zooming into your website.

While text inside images can present a challenge to any visitor, text inside images is especially difficult for visitors with disabilities. For instance, visitors who are blind rely on screen readers to access your website and screen readers can't read words contained in the image. As a result, these visitors won't be able to understand your images regardless of the image size or zoom level.

If key information—pricing, product features, event dates, and so on—must be provided within an image, alternatives need to be provided for people who cannot see the text in the image either due to a disability or due to resizing the website.[21] One option is to add an alt attribute to each image and the text of the attribute could repeat the text included in the image. This helps visitors relying on screen readers be able to understand what the image contains.

The problem is this alt attribute is contained within the code and is hidden from the visible portion of your website shown in the browser. As a result, it doesn't help people who can see the resized image. To help these visitors, captions can be added, showing additional text above, below, or to the side of the image. But if there is already text within the image, repeating the text in the caption would be redundant and potentially confusing to visitors.

Instead, as the website size scales up or down, larger or smaller sizes of the image can be substituted via responsive image techniques.[22] Visitors who access the website using different sized browsers or resize their browser via zooming then see an image appropriately sized for their screen. This keeps the text within the image legible regardless of the screen size or zoom level the visitor chooses. Of course, a downside to this method is that multiple images need to be created and maintained.

The best option is to keep text separate from images, at least as much as possible. This will almost always allow for the greatest amount of flexibility because text can be resized by visitors considerably easier than images.

• • •
Pop-Ups

Many websites generate leads, sales, and other conversions using pop-ups. The term pop-up is a little tricky to define. While traditional pop-ups open some type of notice in a different window, today's pop-ups generally open within the same window, graying out the rest of the website in the background. Technically, these modern pop-ups are referred to as modal dialog boxes and are also known as lightboxes or overlays. The term pop-up can also be used to describe notification or alert windows. Each term means something slightly different, but all describe a notification that pops into a visitor's view.

The biggest problem with all of these pop-ups is a visitor cannot use the website until dealing with the pop-up. In the words of Jakob Nielsen, a leading usability expert, this makes it a "blunt instrument that hits [visitors] over the head."[23] With a pop-up, people are forced to take an action whether they want to or not. Visitors have to either do what the pop-up suggests, or they have to find a way to close the pop-up. Either choice interferes with a visitor's control over how he or she is using the website.

There are legitimate reasons to use a pop-up and prevent people from taking an action. By showing a warning message that somebody is about to move away from a page and lose information contained within a form on that page, you help visitors stay in control and avoid deleting information they had input into the form. In that instance, visitors may not be aware that clicking away would cause the deletion, so the pop-up saved what could be a costly mistake. For similar reasons, using a pop-up to show confirmation messages can help remind people about the choices they've made, giving them a way to bail out if they have made that choice in error.

In contrast, showing an intrusive pop-up that pressures the visitor into a sale or sign up could take visitors off their chosen path. This is especially true if the pop-up appears immediately upon a visitor's arrival to the website. This pop-up might announce a new special offer or a product announcement—and by including it in a pop-up you help to ensure visitors won't miss this critical information.

The problem is by showing a pop-up immediately upon arrival, the person visiting has not yet had a chance to click or scroll through your website to see what is offered and make his or her own determination if the website contains something relevant to whatever it is he or she wanted.[24] Instead, people are forced to close the intrusive pop-up before continuing along their way to wherever it is they wanted to go.

Other pop-ups are based on the time a visitor has spent or a visitor's activity. For instance, after a visitor has spent thirty seconds on a page, a pop-up might interrupt them to ask if they would like to subscribe to an e-mail newsletter. Other pop-ups might appear after a visitor has scrolled half way down the page they were reading. Unlike the pop-ups triggered upon arrival, visitors have had least had some time to review your website. However, these pop-ups interrupt and distract visitors. At minimum, this is a mild annoyance. For instance, a visitor has to close the pop-up in order to continue reading an article. In other cases, the pop-up might be so obtrusive that visitors simply can't figure out how to close it in order to return to doing whatever it is they were doing.

Of course, some people might follow the pop-up and be sent down an entirely different path than the one they intended to take. It could be argued this might take people to a place they ultimately wanted to go during a visit to your website—they just didn't know they wanted to go there until the pop-up suggested it. If that's true, and the pop-up does guide people somewhere they really wanted to go, then this could be beneficial to your visitors. If it leads to more people converting, this can also be beneficial to your organization.

But this becomes a problem if the pop-up takes control away from a visitor, sending them off their desired path. In these situations, people will leave dissatisfied because they weren't able to complete the task they came to your website to complete. Not surprisingly, studies suggest there can be long-term, negative impacts of using intrusive pop-ups because of the way it makes people perceive your website and your organization.[25]

One of the better and more effective types of pop-up for increasing conversions are those that are triggered when somebody is about to leave your website.[26] These exit-intent pop-ups let you offer one last call to action to people who are about to leave.[27] For instance, if somebody is about to abandon a page discussing a product, the exit-intent pop-up could offer a coupon code if the person makes the purchase right now.[28] Some people may be tempted enough by this offer that they choose to stay on your website and convert. The more you can identify the reasons people leave, and the offers that might encourage them to stay, the more effective this style of pop-up will be.

Like all pop-ups, exit-intent pop-ups are still obtrusive because they block visitors from doing what they wanted to do (in this case, leaving). But the impact on visitors in lessened somewhat because you didn't block them from doing something during their visit. They could still move around your website, engaging and reading your text, without pop-ups interfering. Instead, when people leave, the exit-intent pop-ups offer you one last way of helping a visitor get something they potentially wanted. The worse that can happen is people ignore the exit-intent pop-up and leave, but visitors who saw these pop-ups were leaving anyway.

Behavioral Considerations

• • •

Choice Overload

Part of changing your website includes adding new pages that provide more details about everything your organization offers. In many cases, you will have lots of supporting pages to further explain why people ought to work with and care about your organization. Along with adding value to your organization, these additional pages also are valuable for your visitors because each additional page offers one more option for people to choose during their visit.

But too many options can potentially overwhelm people, causing anxiety.[29] When too many options are presented, people are not always sure which option they should choose—"too many" is a relative amount, varying based on who your visitors are and the different nature of the choices offered. Overwhelmed by all these choices, visitors are unable to decide and instead opt to leave the website without making any choice. This has obvious negative impacts on conversions and engagements.

One way to interpret this research is to conclude there is a problem from offering too many choices. That conclusion has helped lead to more research suggesting limited choices can lead to higher conversion rates.[30] There is evidence of this reduction of choices increasing sales offline as well.[31] By removing options, you help prevent people from feeling anxious or overwhelmed, lessening the chance people will leave. By removing all options but one, you end up with more people choosing the one remaining option—if that one remaining option is a way for people to convert, this will likely increase conversion rates.

But it's also true that people can be frustrated by the lack of options, and this lack of options can hurt your conversions too.

People have unique needs and expectations, so when they visit your website, they want to find what is right for them. Do you want to risk losing conversions because you stopped featuring a product some percentage of your visitors were interested in purchasing? People don't want to be forced into one choice any more than they want to be overwhelmed by too many choices.

What all of the research together suggests, though, is this isn't about how many options to offer, but rather the way those choices are offered. As Jason Fried of 37Signals notes, "It's not about ten features versus seven, it's about the right four versus the wrong eight."[32] In other words, it's fine if you have many options to offer—people may actually prefer having so many options, provided each option is of interest. After all, if you limit choices, people might leave because the remaining options aren't what they wanted.

This perceived choice overload has more to do with how those options are organized and presented to your visitors. For instance, Google provides millions of results to most search queries. This large volume could certainly be considered overwhelming and anxiety inducing, leading to a difficult choice for people to make. But Google search results typically aren't overwhelming and don't produce anxiety. Instead, Google's search results are organized so that you rarely need to look beyond the first handful of results returned to find what you are seeking. Chris Anderson, editor-in-chief of *Wired*, summarized this well, saying, "Order it wrong and choice is oppressive; order it right and it's liberating."[33]

Too many options or a handful of options poorly presented can make people feel overwhelmed, causing them to leave your website without engaging or converting. To help your visitors and to ensure they are satisfied with their visit, you need to offer several options for them to choose between. But these options need to be organized so people can easily choose the option that will best meet their needs.

This means part of optimizing your website is deciding what choices to offer where and how to present those choices. What choices to offer will differ across the various pages of your website since people have different expectations when they arrive on each page. You want to try different quantities and types of options, along with testing different ways of presenting those options. You need to determine what number of choices and what presentation style gets your visitors engaging more deeply and converting more frequently.

For example, a complex page intended for people conducting research may need to offer several options—like links to resources like videos, images, graphs, charts, or PDF files. All of these options will help them conduct their research and answer the questions they are asking. If you organize these links by how in-depth, trustworthy, or recent the resources are, you can help people make more informed decisions about which resource they should access.

However, a page encouraging people to take a specific action—like purchasing a product—would need fewer options. If people arrive on a page expecting to purchase a product, promoting anything other than a Buy Now option would get in their way of making that purchase. For visitors who arrive on this page and don't want to purchase the product, the product page could also provide an option to learn more in order to not force people into making a purchase.

• • •

Information Foraging

The theory of information foraging was proposed in 1999 as a way to describe how people hunt for information.[34] Its roots are in anthropological and ecological explanations of how animals hunt and search for food. The food foraging theory that the information foraging theory is based on describes how animals seek to conserve their energy while hunting prey. By selectively choosing their prey, animals avoid wasting energy chasing prey they have little chance to capture and eat.

This theory of the animal kingdom translates to how people

browse the web: we want to browse the web, clicking one link after another, scrolling page after page, exerting as little effort as possible in the process of finding whatever it is we want to find.[35] To offer the most freedom and control to your visitors, your website should allow people to hunt for what they are seeking while exerting the least amount of energy. If your website appears difficult to use—by showing too many disorderly options or by pressuring visitors into converting—too much energy is required. People will leave without completing their hunt (maybe to hunt on a competitor's website).

Animals need cues while foraging for food, and one of the best cues is scent. A similar concept can be seen with people visiting websites where text and design features provide visual cues, giving them an indication of what type information exists on a website they are currently visiting. The more people find cues telling them your website contains what they are after, the more likely it is they will stay on your website. If your website lacks these cues, people will seek out their desired information elsewhere even if your website did contain the information they were seeking.

One of the primary ways of providing these cues to your visitors is through your website's navigation. This is why the words in the navigation need to speak the language your visitors use. The more the words match how people visiting think and talk about what your website offers, the better the navigation will work at providing cues. The better the cues, the more people will click or tap on the navigation to browse deeper into your website. Similarly, the navigation should be visually designed to draw people's attention so that visitors know to look at and use the navigation.

This theory also helps to explain why people visiting a website from a smartphone aren't currently using the hamburger icon to access the website's navigation. People aren't sure what the icon is, making the icon an ineffective cue. By adding the word Menu or putting a box around the icon to make it look like a button, you make the cue clearer to people so they know where the navigation is located and how they should use it.[36]

What to Measure

• • •

Visitor Intent and Meeting Expectations

Offering guidance while still letting people control their visit requires finding the right number of options to provide along with presenting those options correctly. To help determine what options to include and what to exclude, you need to know what visitors expect and what they intend to do when visiting any particular page. By knowing visitor expectations and intentions, you can decide what text, features, calls to action, or links to other pages should be included on a particular page.

One way to know what people expect and intend is to ask them directly, for instance by using surveys or interviews. If you have the means to do so, surveys and interviews are a great way to understand what people think and expect. But to get accurate and reliable information, you typically need to interview or survey a wide group of people. As well, you need to ensure the profile of the sample group interviewed or surveyed matches the profile of actual visitors to your website. As a result, running an effective survey or conducting useful interviews can be difficult if your organization has a smaller budget, a short amount of time, or a small base of customers to recruit to your survey. Surveys and interviews are also challenging for new companies that are not yet clear on what audience they will target.

In lieu of surveys or interviews, you can start to understand expectations by reviewing the sources people use to reach your website or for new websites, reviewing the sources people use to reach competitor or other similar websites. For example, if many people arrive on the website after using a search engine, like Google, you want to know what search phrases were used. Once you know the phrases used, the phrase will give you an idea of people conducting that search are expecting to find if they were to click to your website.

If people search for the term "what are blue widgets," they probably want to learn what blue widgets are, but aren't as interested in making a purchase. Given this, the page people find on your website following this search should provide information to meet those more educationally minded expectations instead of pushing immediately toward a sale. If people search for "buy blue widgets," they are interested in purchasing blue widgets and the page people reach on your website after this search should help facilitate that sale—by pushing toward a conversion.

A similar process can help you gauge expectations for people arriving from other sources. Another source leading people toward your website might be other websites linking to certain pages or blog posts. By reviewing these other websites, you can determine in what way they are linking to you by reading the rest of the page—is this other website promoting your organization positively or negatively, or are they reviewing a product or a service your organization provides? All of this helps to explain what people are thinking when they click from that other website to yours. Similarly, you can review what people have shared about your website and your organization on social media, local listing websites, or forums.

Using traffic sources to judge expectations is educated guesswork, but it gives you a place to start writing and organizing your website's text so that it matches visitor expectations. If you know what people are thinking when they conduct a search, or you know what they have recently read on a page linking to your website, or if you know what discussions reference your organization on social media, you can make an educated guess about what people expect from your website and what kinds of things they'd intend to find upon visiting. This changes the options you'll provide to those visitors.

As well, by reviewing these sources leading people toward your website, you can understand the right order in which to present these options. You may find that people arriving on your website are more likely to be interested in finding information about the seminars your organization hosts each month instead of information about becoming a volunteer for your organization. While options to learn about both can be provided, because more people are interested in the seminars than the volunteer work, more links can be provided to help people locate this information.

The more your website's text, links, images, design features, functionality, organizational structure, and other items represent what your visitors think, the more your website will be able to meet people's expectations and intentions. This allows more people to feel like they got what they want, how they want, and when they want during their visit.

$$\bullet \quad \bullet \quad \bullet$$

Bounce Rate and Distraction Rate

As you get visitors to your website, you need to confirm the educated guesses you made about visitor expectations and intentions. Educated guesses based on visit sources give you a way to begin understanding expectations, but you can refine these guesses based on how people respond to the changes you make to the options offered. One of the more helpful metrics to gauge if the pages on your website meet people's expectations is the bounce rate.[37] The bounce rate is a measure of how many people came to your website, looked at one page, and then left without doing anything beyond looking at that first page.

If you expected people to come to a page on your website and then continue to another page immediately afterward, a high bounce rate suggests the various items on your website are not delivering what people want and giving them a reason to click to that next page. Without finding something they want, people will bounce away instead of engaging or converting.

A higher bounce rate usually means you should dig deeper to understand why your page isn't meeting visitor expectations. Perhaps the text needs to be rewritten with different calls to action or links to other pages should be included. You may also want to review the sources with the higher bounce rates—visitors arriving from a Google search might have a low bounce rate, but people arriving from an advertisement in the local newspaper might have a considerably high bounce rate. Seeing the higher bounce rate for one source would suggest the page people are arriving on from that source, the advertisement in this example, isn't meeting their expectations—directing people to a different page or adjusting the source itself may help to address this high bounce rate.

Alternatively, a high bounce rate may not be a problem. This is especially true for pages where people seek quick answers. In these cases, one page may be enough for people to get the information they needed. People may want to find your store's address, so they clicked a link from some source and arrived on your contact page. After arriving, the visitor retrieved the address and left your website perfectly satisfied even though they never clicked to another page. In this example, anything other than a high bounce rate would be a greater cause for concern.

Before worrying about a high or low bounce rate, you need to know how people want to engage and interact when reaching each page. Some pages will have naturally lower or higher bounce rates than others. Once you know what type of bounce rate you'd expect from each page, you can determine what a good bounce rate is for each page. If the bounce rate for a page differs from your expectations, either by being too high or too low, then you can worry and work on making appropriate changes.

You can also clarify visitor expectations and intentions by measuring each page's distraction rate.[38] Distracted visitors are, by definition, not engaged, and if they are too distracted, they may not convert either. There are two parts to measuring the distraction rate: your engagement or conversion rate and the time spent on a page or on the website.

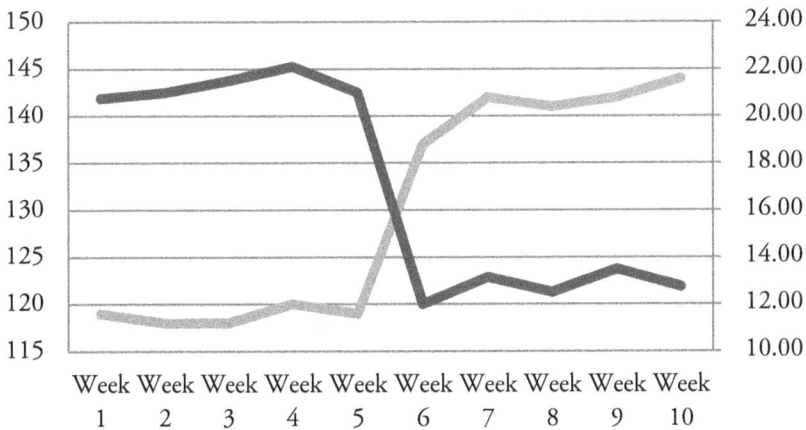

Figure 2-2. Graph of distraction rate. Light gray line shows an increase in time on website starting in week 6 with time shown in seconds on left axis. Dark gray line shows a decrease in the engagement rate starting during week 6 with engagement rate shown on the right axis. Whatever changed during week six distracted visitors.

If after making changes the time spent on a page increases but conversions or engagements remain flat or decline, as shown in figure 2-2, people are probably distracted as a result of those changes. That distraction might come from an advertisement or pop-up taking people's attention away from the engagement or conversion they wanted to complete.

Alternatively, the distraction might result from unclear options presented in your navigation or links within the text. The more people are distracted, the more time they have to spend trying to figure out how to do whatever it is they wanted to do instead of actually doing whatever it is they came to your website to do. This leads to greater frustration and less satisfaction during a visit to your website, which explains the decreased engagements and conversions.

* * *

Visitor Journeys and Paths

Another key aspect to measure is how people move throughout your website. Knowing the various ways people get from point A to point B will help you decide how to adjust your website's text and design. The point B in this measurement is just some endpoint of a path people took. It can be something like a page reached after placing an order, or it could be a key page you want people to access when they are visiting your website, like your contact page or a page listing different types of products.

Whatever the endpoint, you want to find the current paths people take to reach it. Where problems exist along that path, like high bounce or distraction rates or low conversion or engagement rates, you can adjust the text, images, links, navigational structures, design features, or other items to help get more people toward their desired endpoint. Most analytics tools can show you the various ways people moved through your website, and these paths can be filtered to only include people who reached a specific endpoint.

As you start to measure the various paths people take, the data can seem overwhelming given the number of paths that can exist. On websites that provide high degrees of control and freedom, there are many different ways people can move through a website, clicking on top navigation, sidebar navigation, in-text links, calls to action, submitting forms to move to a new page, and on the list goes. Think of it this way: if you have a six-page website where people have the option to move from one page to any others, there could be 720 ways in which those six pages could be browsed through. A twelve-page website with options to move to any one of those pages has over 479 million ways for people to browse through.

While the paths may prove overwhelming and the data may seem large and messy, this large amount of paths represents a great deal of freedom and control for your visitors. A website with fewer paths for visitors to take would have simpler data to review, but would offer people less control over their visit. Of course, if you are trying to take control away, in the case of landing pages or an order process, you may only want to see one path within your analytics tool.

Ideally, you could track each link in every path from every visitor. With so many possible pathways for people to take, it isn't practical to track each one. Instead, you want to identify the most common paths people take to reach the various endpoints. This will help you figure out what links, words, design features, and pages people use as they hunt for information on your website. To offer more guidance to your visitors and to get your website thinking like they think, you can use more of the links, words, and design features people actually use to navigate the website while dropping the words or features people aren't using.

Some of the more helpful paths to track are the pages people went to immediately before they completed a conversion or engagement. Knowing the pages people were on before they converted or engaged gives you a good idea of what led up to that conversion or engagement. For instance, you may find many people who added an item to their cart on page C were previously on page B. Before they were on page B, almost all of these people were also on page A. This pattern suggests people who go from page A to page B are quite interested in the product displayed on page C. In this example, to increase the people clicking the Add to Cart button, you may want to consider adding links to page C on page A so that people don't have to waste their time going to page B first.

Reviewing reports about paths can often be more helpful for what is not shown: if you look closely, you'll often find paths you expected people to take that are simply missing. There might be links people are not clicking, or there may be entire pages nobody ever visits. This could indicate you should remove those unclicked links or unvisited pages. Removing would allow room for a more prominent promotion of the pages people do visit and the links people click. Of course, missing paths can also indicate an opportunity to rework how links are worded—perhaps people aren't clicking because the words in the link don't offer a good cue for why they should click to that page.

When reviewing the paths people take or don't take, it's quite likely you'll find confusing paths that make you wonder why people moved through your website in a certain way. Chances are, people don't and won't move through your website the way you move through your website. That's precisely the point of giving people control over their visit to your website: people should be able to take many different paths. Given that, if something confuses you, that doesn't necessarily mean it's a bad path as it may be perfectly clear to your visitors.

For these confusing paths, you'll need to determine if it's a confusing path to just you or also confusing to your visitors. If the paths end up with people reaching a desired endpoint, or if people who take that path engage and convert, then it's likely the path made some kind of sense to the visitor. There may be ways to simplify the path or make it more efficient (as discussed in chapter 1), but this path is proving beneficial to your visitors. If, however, the visitors who took the confusing path never reached a target endpoint or never engaged and converted, then the pages and links along the path should be revised. By adjusting the options offered or the ways those options are worded, you could reduce at least some of the confusion visitors are no doubt experiencing.

Supporting Measures

• • •

Screen Resolution and Viewport

Even if the choices offered are well organized, there is a limit to how much can be shown on each page given the visitor's device. Every piece of text, image, navigation, call to action, or design feature takes up on space on the visitor's screen. The available space is best measured by screen resolution. Somewhat oversimplified, but the bigger the screen, the more room you have available to fit in navigation, calls to action, text, images, videos, and other items.[39] On smartphones or tablets, there is less room available on the visitor's screen than on laptops or desktop computers.

Smartphone screen resolutions are usually around 360 to 568 pixels wide and 320 to 640 pixels tall, though larger (and smaller) resolutions exist.[40] Some tablets have the same resolution as smartphones but larger tablets extend to 1024 to 1280 pixels wide and 600 to 800 pixels tall.[41] Desktop computers have a wider range, extending to over 2000 pixels wide and over 1000 pixels in height.[42] See table 2-1 for examples of common sizes.

Desktop/Laptop	Tablet	Smartphone
1366×768	768×1024	360×640
1920×1080	1280×800	375×667
1440×900	600×1024	320×568
1280×800	601×962	480×800
1280×1024	1024×768	320×534

Table 2-1. Common screen resolutions per device. Data from StatCounter's Global Stats. Note that browser viewports will be smaller.

There are, of course, complications in relying on screen resolution information. Smartphones and tablets can be oriented horizontally or vertically, altering the resolution and changing how much of your website visitors can see.[43] The bigger complication though is that browsers add in menu bars, address bars, and tool bars, further limiting how much of your website can be seen. So, even if a visitor has a screen resolution with a height of 1024 pixels, the amount of that available to view your website might only be 824 pixels because 200 pixels are used by a browser's menu bar, tool bar, and address bar. The space available within the browser window is referred to as the browser's viewport.

There is a lot of variety in screen resolutions and even more variety in viewports. Undoubtedly, new devices with new resolutions will continue to appear. You don't need to worry about every available screen resolution or viewport, but you need to know the common screen resolutions and viewports used by your visitors. This information will help you alter your design to ensure critical items are placed where visitors can see them.

One way to identify areas your design needs to change is to review conversion and engagement rates for each resolution and viewport— if some widths or heights have lower rates, adjustments can be made to your design to better support those visitors. Conversions may simply be lower at a particular viewport because people are unable to see a call to action button—visitors want to convert, but your website's design makes doing so difficult or impossible. Also, some resolutions with wider widths might spread elements out too far across the width of the page making it challenging to skim through your website's text.

One key area to pay attention to is how your text changes based on different screen resolutions. On smaller resolutions, the text might be crowded in alongside all the other features. On larger resolutions, there may be too much text on each line. Either way, this can make the text uncomfortable to read. The more challenging it is to read the text, the more people will struggle to understand what your website offers and how they should navigate through your website. On larger screens, sixty to seventy-five characters per line allows for comfortable reading and on smaller screens, the font size and design should change to only show thirty-five to forty characters per line.[44]

• • •

Heatmaps and Visual Recordings

Part of allowing people greater control over their visit is ensuring links and navigation options not only include the words or phrases they expect, but also that these options are placed where people expect to find them. To support this expectation, you want to place the things people expect to find where people are looking. One way to know where people look on a page is to use eye tracking devices, measuring where a visitor's eye gazes or moves. These devices are expensive and require the visitor wear a particular device to monitor eye movement. While useful in a laboratory, for day-to-day changes, a simpler solution is a heatmap.

Heatmaps provide a visual report on where people have clicked or tapped on a page. The click information allows you to know what links, buttons, navigation, or other clickable parts of a page draw people's attention the most. Relying on this information can help you decide what changes to make to offer more guidance to your visitors. Heatmaps also report on scrolling activity, as discussed in chapter 1. Scroll activity offers another way to determine a visitor's viewport size since it tells you how much of your website they can see before having to scroll.

To an extent, heatmaps also give you an approximation of where people are looking on your website because heatmaps can report on mouse movements. Some studies have found the mouse and eyes move together up to 84 percent of the time and other studies have found this correlation is strongest when visitors are doing things like reading a page or actively completing a task.[45] You can see an example of a mouse movement heatmap in the top portion of figure 2-3. The brighter areas of color indicate areas with more mouse movement, which suggest more visitors are looking at and engaging with these areas.

Mouse movements are not a perfect correlation and won't apply to all situations. The biggest problem is that mouse movement heatmaps are obviously unavailable for people visiting your website from mouseless devices like smartphones or tablets. But, despite these drawbacks, mouse movement information can provide a means of approximating where people look, giving you an idea of where to place key information on your pages.

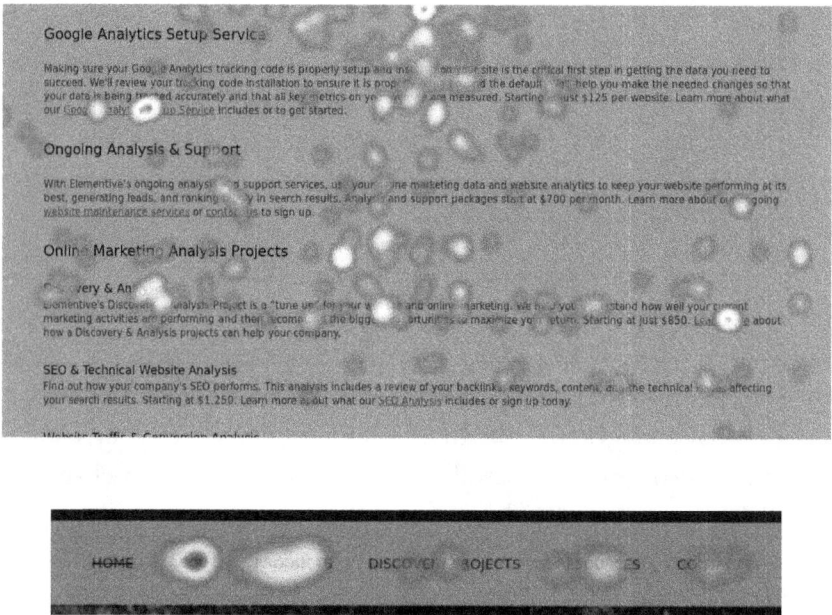

Figure 2-3. Top: mouse movement activity as reported within a heatmap. Bottom: Click and tap activity as reported in a heatmap. Brighter areas indicate more movement, clicks, or taps.

Another way of monitoring how people navigate your website is by recording your visitors. This lets you watch the person's entire visit—every link clicked, every scroll, every mouse movement and more. Visual recordings give you a clear order of the paths taken by your visitors. However, getting the most out of visual recordings requires watching the playback of many visits to detect patterns in the ways people navigate your website. Because watching so many recordings is time-consuming, this measurement technique is best used to review visitor paths on a semiannual basis or before making major changes that will affect those paths.

Reviewing heatmaps or watching recordings can help you find the optimal number of choices to offer and the optimal way to present those choices. If on a heatmap there is a great deal of scrolling, movement, or attention in a certain area of a page but no clicks or taps occur in this area, visitors might be confused about what to do. Adjusting the link or button text in that area of the page or reducing the number of links or buttons might ease the confusion, letting people have an easier time of choosing where they want to go next.

Evaluate Your Website

Organizing Navigation and Offering Guidance

- Do the pages included in the navigation represent topics of interest to your visitors or topics your organization is interested in making people look at?

- Are links in the navigation and links included elsewhere on your website written to provide clear cues for what people will find if they were to click or tap on that link?

- Is there some form of navigation on every page of your website? Is the navigation obviously placed so people can easily locate it?

- Are too many options provided within the navigation or too many calls to action included on any page? What navigation links or calls to action are not being used and could be removed?

- How are the links in the navigation organized? Does this organization structure make sense to your visitors?

- Are visitors engaging with the navigation currently? If not, how could the links be reorganized to encourage more visitors to engage and utilize the links in the navigation to browse through your website?

Allowing More Control

- Can people move freely from one part of the website to another? What limitations does your website place on visitor movement? Are these limitations justified?

- Are there features on your website forcing people to take a certain action? How will these features support visitors who are uninterested in taking that action?

- Before canceling any actions of high consequence—like removing a product from a cart or abandoning a partially complete form—are people prompted to confirm the cancellation? Are methods provided to help people undo the cancellation, either on your website or offline?

- Do pop-ups block people from doing something they want to do? Within the pop-up, are visitors given a chance to easily bypass the pop-up and resume what they were previously doing?

- If key information or text is included inside of images, are alternative means of interacting with that image provided?

Measurement Guide

For guidance setting up the tools for these measurements, see:
http://www.matthewedgar.net/elements/control

Baseline Setup	Identify the various paths you would expect people visiting your website to take. Identify the endpoints you would expect people to reach after taking a particular path, like a confirmation page people might see after converting.
	Setup heatmaps on key pages to see how people click, scroll, or move through the page. Use this to find changes to make and evaluate the impact of those changes.
Monthly or Quarterly	Review the paths people actually take on your website and the various endpoints they reach. As you compare this to the paths you expected to take, note where your expectations were right or wrong—this will help you learn what people want to do during a visit. Also, identify the most common paths taken. What changes would make these common paths easier or simpler?
	Monitor bounce and distraction rates from the various sources leading people to your website. How can you improve your website to help people arriving from sources with higher bounce or distraction rates?
Before Major Changes	Monitor bounce rate and distraction rate along each path along with conversion and engagement rates. If any of these rates show negative gains after making a change, that change probably didn't help.
	Review the screen resolutions visitors are using to access your website on different types of devices. View your website from these resolutions—what is missing, and what problems exist at that resolution?

CHAPTER THREE

MAINTAIN CONSISTENCY AND FOLLOW STANDARDS

SIMILAR THINGS SHOULD behave in similar ways. This is a basic expectation of websites and life in general. If you are using an object that looks like an object you've used before, it's reasonable to assume it should work the same way. Whether dialing a phone, opening a door, driving a car, or ordering something on a website, there is an accepted norm for how the various objects used to complete those tasks should work.

If things don't work consistently with expectations, this causes a sense of disconnect due to unfamiliarity, leading to confusion. Imagine if every car's steering wheel operated uniquely or if every phone organized numbers on the dial pad randomly. Like all new technologies, early websites rarely followed standards or behaved consistently. As websites have matured, standards have formed, making today's websites far more consistent than in the past. Some websites, though, still defy the standards (rarely to their benefit).

To avoid confusing visitors and causing a sense of disconnect, your website needs to follow both external and internal standards.[1]

- Maintaining external consistency requires following best practices and accepted conventions. These are the standards of websites in general but also the standards of websites in your industry. By following external standards, your website will behave similar to others allowing people to feel comfortable using your website.

- Internal consistency requires following your own conventions for how everything on your website should look and behave. This means each page, line of text, image or other design features should look like it all belongs together. You don't want to annoy visitors as they waste their time figuring out how each part of your website works.

Key Concepts and Questions

• • •

External Consistency

When visiting a website for the first time, the general assumption is the website will behave like all of the others people previously visited. Your visitors are typically not interested in investing their time in learning how to use your website. Instead, they want to use your website to get what they need as quickly as possible. In order to move as quickly as possible, visitors will rely on past experiences to determine how your website should work. If your website behaves in unexpected or unique ways, this will slow people down, making your website more difficult for them to use.

For example, when using a website on a desktop or laptop computer, people assume the logo will be located in the upper left corner, usually providing a link back to the home page. According to some studies, placing the logo in another location makes it harder for people to remember the website they visited, which has implications for your organization's brand recognition and long-term success.[2] Part of the reason this standard exists is because most people focus more on the left side of the screen (or, if your visitors read right-to-left, the reverse will more likely be true).[3]

Why people focus on the top-left corner, at least for people whose predominate language reads left-to-right, is a chicken-or-egg question. It's possibly because people who read left-to-right find that looking for the logo on the left feels more natural. Or it's because the convention has been well established across websites and people know to pay more attention to the left side of the screen when looking for things like the organization's logo. Either way, by not including the logo on the upper left corner, your website will be inconsistent with what others have done, and this deviation will make people have to take longer figuring out how to use your website.[4]

Along with maintaining consistency with the web overall, you also need to maintain consistency within your industry. When evaluating your products or services, people will evaluate those same products or services on competitor websites—no matter how much you'd prefer otherwise. As they browse through other similar websites, people will expect each website to show the same type of information—after all, to the visitor, you and your competitors are similar entities. Within any industry, there is a standard way to show pricing, photos, testimonials, demos, blog posts, contact information, products, services, events, donation requests, or more. The more you include these similar types of information and the more visually similar they appear, the more you will demonstrate to visitors that you are a viable option within your industry. As well, the more similarities, the easier it will be for visitors to compare you to your competitors, allowing them to see why your organization is, obviously, superior.

Maintaining consistency with other websites in your industry will also make your website easier to use. When visiting your website, visitors can rely on past experiences from other websites in your industry and find the information they are seeking more easily. If your website relies on standards, people visiting won't have to waste time figuring out what makes using your website unique. Instead, they will be able to invest their time engaging with your website to learn more about why your organization is unique. This allows your website's technology and design to get out of the way, letting people stay focused on what matters most.

Of course, if your website looks like every other website, it becomes forgettable. When your website looks almost exactly like your competitors' websites, people will know how to use your website, making for a very simple visit. But they will probably confuse you with your competitors, which has significant drawbacks. As you change your website, you have to decide where to deviate from best practices and industry standards.

One deviation designers have recently experimented with is

placing the navigation on the desktop version of a website along the bottom of the screen instead of the more conventional top of the screen placement. This has had limited success, tending to work best on one-page websites for technology-oriented audiences.[5] This new design requires people to learn a new location for navigation. Tech-savvy audiences might be more willing to spend time understanding about this location than other audiences. For better or worse, most people visiting a website from a desktop computer are familiar with the navigation near the top of the screen.

As you decide where to be inconsistent and deviate from accepted standards, you want to do so for a good reason and ensure people don't have to learn new ways to use a website or learn new ways to use different parts of your website.[6] The best recommendation is to use conventions and industry standards as the starting point. From there, you can see how people use your website and then slowly make modifications that make different pages or parts of your website unique and more memorable to distinguish you from other organizations. As you make changes, keep an eye on whether the deviations are helping or hurting conversion and engagement rates.

As an example, many business-to-business websites struggle to decide whether to show pricing. Being one of the few websites in your industry to show pricing might give you an advantage over competitors, letting you have more influence over potential customers' thoughts as they make a purchase decision.[7] You violated industry standards and are inconsistent, but in a way that benefits visitors and, potentially, your organization. While everybody else hid the price, your organization didn't. This gives people a reason to engage with your website, remember who you are, and, eventually, could be part of the reason people convert.

Of course, if your product or service is heavily customized, showing pricing information while others don't might prove confusing to visitors. This confusion could drive people to a competitor's website where the confusing pricing information is kept hidden from view.[8] In this case, choosing to deviate from standards and show pricing information was a disadvantage to your visitors and your organization.

Let's take this same example from the reverse. If most competitors show pricing information, following this industry convention could make your website look like everybody else's, potentially leading to visitors confusing your organization with your competitors. Along with leading to your organization being forgotten, this may also work against your organization's interest if your product or service isn't like everybody else's. Keeping the pricing from view or making people contact you for pricing details could set your website and organization apart. This could present your organization as higher-end or more exclusive, differentiating you from everybody else.

However, not showing pricing information may frustrate your visitors, especially if they have to take an extra step of contacting your organization to access pricing—none of your competitors wasted their precious time by making them take this step. To succeed with a deviation that increases the steps a visitor is required to take, your website needs to offer a compelling reason why your organization is worth the extra effort. For instance, if you make people call you to access pricing, your website's text could explain how much better your products or services are or your website's design could show how high-end your organization is. You can also explain to visitors how they personally will benefit from this extra step of contacting you for pricing, such as customized services or personalized attention.

Deviating from industry standards comes with a risk, but so does close adherence to the standards. To prevent your website from becoming forgettable, differentiation is critical. Because these differences can make your website more challenging for visitors to use, you want to measure how these deviations from the expected norm work for your visitors. Keep the deviations that increase engagements, conversions, or make your website more memorable. But remove any deviations from the expected norm that prove frustrating or confusing.

• • •

Internal Consistency

Within your website, you have the opportunity to define the rules for how things work. This gives you immense control over making sure everything on your website looks and behaves in similar ways. Despite the control you have to enforce internal standards, there are many forces pulling each page, feature, image, and paragraph in inconsistent directions.

One of the larger challenges in maintaining internal consistency comes from how many aspects there are to your website. When considering fonts, you need to know: What font family is used? What size? Will tables or graphs use the same font family and size or something different? What words are bolded or italicized? What style will you use for headers or subheaders? Will there be subheaders? What about sub-subheaders? What colors will you allow for each type of text? Will images have captions and, if so, what style of text will those use?

Along with font considerations, the tone, cadence, and structure of the text need to be consistent. You also need to be consistent with the use of various design features—like buttons, form fields, tables, links, images, charts, graphs, video containers, itemized lists, or more. In addition, every aspect must work with the next—a website using a soft, educational tone within the text would be inconsistent if it used a loud, garish visual style.

After deciding what the standards are, the next challenging part is enforcing these standards. Say, for instance, you need to add a new page to your website—how do you keep the text on this new page consistent with the text on other, older pages? The original writer of your website carefully wrote each page following clear standards. The tone and cadence were similar between pages if for no other reason than the same person wrote it over a short time period.

The problem arises if the original writer is no longer available when you need to add new pages or edit existing pages. Within the new text, the tone shifts, breaking the consistency. Depending on how far the tone shifts, this inconsistency might be a minor nuisance and barely noticeable, or it may present a major source of confusion or dissatisfaction with your website.

Design presents a similar type of challenge when adding a new feature to your website. This new feature may require a different design than anything currently on your website—such as a new table, fields in a form, a new style for a call to action, or a way of displaying photos. It's impossible for the original designer of your website to know every feature you will eventually need to create. It's also unlikely the original designer will always be available to work on these new designs. Like with writing new text, as you add new design features, you need to ensure they are consistent with everything else to prevent causing any problems for the people visiting, even if the original designer is no longer available.

●   ●   ●

Device Consistency

As a part of maintaining consistency, you also want to consider the consistency of your website between all the different devices your visitors can use—desktop computers, laptops, tablets, smartphones, and, depending on your product or service, televisions. A visitor to your website may start out looking for information on his or her phone or tablet, but move from the phone to a desktop or laptop computer to finalize the sale.[9] People working on complex activities—like planning a trip or managing finances—are more likely to start their search on a desktop computer before continuing the search on a smartphone.[10] But people shopping might start on a tablet before completing the purchase on a desktop computer.[11]

When people visit your website from multiple devices, they have an expectation that your website should behave similarly and contain the same information regardless of the device used. If your website shows pricing or a detailed list of features when accessed on a desktop computer, then the same pricing and list of features should be available when visiting from a smartphone. It's tempting as you scale your website down to smaller screen resolutions to show less information—after all, less stuff can fit on a smartphone's screen than on a desktop computer's screen.

However, if the information you present is inconsistent between different devices, visitors will struggle to understand how to use your website as they move from one device to the next. Effectively, by removing something from your website depending on the type of device a visitor used, you are penalizing people for choosing to visit your website on that certain type of device.[12] The more people wanted whatever was removed from the smaller-screened version of your website, the more this penalty will hurt the long-term success of your website and, quite possibly, your organization.

This is true of information and features, but also applies to the visual style of your website. People visiting your website rely on your design to help them remember who your organization is. If your smartphone website is full of pictures with a modern-design aesthetic, but your desktop website is text-heavy with a design that looks ten years old, people accessing your website from these different devices will have a harder time determining if they have reached the same website. If your website doesn't look familiar, people will probably leave assuming they've accessed the wrong website. The more consistent your website's visual look is across devices, the more familiar your website will appear to people visiting via multiple devices. This consistency will increase the chances of more cross-device conversions and engagements.

Some inconsistencies will naturally exist in how your website is displayed on different devices. Navigation designed for a large screen won't fit or work well on smaller screens, so a different type of navigation is needed for each type of device and screen resolution.

But different navigation styles can still contain links to the same pages so that people can access whatever pages and information they want to find regardless of the device used.

Inconsistencies like the unique navigation per device fall more into the realm of external consistencies, or best practices, for how to design for a given device. Following best practices, overall web standards, and standards within your industry can help to ensure your website's design is modified correctly for various devices. But despite device-specific modifications to help improve usage, people should be able to tell that your website is your website. You might have larger buttons on the design as presented on desktop computers than on the design as presented on smartphones, but those buttons should have similar wording, similar coloring, and when clicked or tapped, the button should take people to the same place.

As much as you should try to keep your website consistent across devices, you do sometimes have to remove text, images, functionality, or design features as you scale to smaller sizes. It isn't always practical to fit every bit of text and every feature offered on the desktop website onto a smaller screen. If something has to removed, a method should be offered allowing people the ability to return to the full website from their smaller-screened device. That way people still have a way to get whatever it is they wanted during their visit.

• • •

Maintaining Consistency

The key to maintaining consistency is establishing standards for how your website will work—what best practices will you follow, what industry conventions are to be adhered to, and what internal rules will guide the development, design, copywriting, and ongoing changes of each page. There are two problems in establishing standards for your websites: first, knowing what the standards should be and, second, getting people to follow these standards.

In some organizations—big or small—the standards are based on what the person in charge of the website prefers. If the person in charge happens to like yellow buttons, all the buttons will be yellow. If the person in charge is continually involved with the website, reliance on personal preference can lead to consistency—assuming that person's tastes don't change.

Of course, relying on personal preference might make your website consistent, but it won't guarantee the strongest engagement or conversion rates. While the person in charge may like yellow buttons and think they look good, your visitors might not agree. In fact, you might find visitors engage and convert more when the buttons are blue. You want consistency, but you want consistency that gets results.

Instead of relying on personal preferences or opinions about how to standardize your website, you need to review what design features, functionality, text, or organizational structures have proven to get more visitors to engage or convert. As one example, pages with photos showing your staff working in the community might get more people to donate to your nonprofit and subscribe to your newsletter than pages with generic stock photos. So in that example, to keep your website's pages consistent while also getting better results, all of your photos should show your staff working in the community your nonprofit serves.

Once you have established how to make your website consistent, the next challenge is getting people to remember to use those standards. Often, website standards are tacit knowledge, stored in the heads of the one or two people who regularly work on the website. The less frequent the changes, however, the harder it becomes for those one or two people to remember each standard. If you are responsible for adding twelve new images to your website each day, it will be easier to remember the standards for size, shape, and style of those images simply because of the repetition. But if you only update images once a year, your memory of those standards will naturally be a bit fuzzy and your website will become increasingly inconsistent.

Tacit knowledge also falters when more people are responsible for updating your website. Your designer remembers the visual standards, your developer knows how to keep the functionality operating the same, and the copywriter remembers the specific tone or style to use. However, what happens if one of those contractors or employees resigns from your organization? Often, this leads to a mad dash to pass the knowledge in the head of the person leaving into the head of another employee or contractor who hasn't left (yet).

The common solution is to preserve the standards in a formal document people can refer to when updating the website. Putting this knowledge about standards to follow into written form allows new and old employees or contractors to access it, lessening the risk of inconsistency when somebody quits. But every standard can't or won't be explicitly stated. Some things are omitted as an oversight. Others are deemed so obvious that their inclusion in the documentation seems useless—until months later when your website's pages are designed in a dozen different ways.

The bigger problem with documentation, though, is getting people to rely on it when making updates. It's tempting for the person updating the website to skip the notes—after all, looking up how to crop the image to conform to the right size, shape, and style takes time, and this image needs to be cropped right now so it can be sent out in the weekly e-mail newsletter in two minutes.

Documentation also can't predict the future. Maybe you need to write text and create images for a new feature, meaning existing standards don't seem to apply—at least not directly. The more situations like this, the more the documentation looks like an unhelpful, archaic relic. As the documentation grows dusty, your website grows inconsistent, and your visitors will grow confused about how to use your disorderly website.

The better solution tends to be keeping standards in written *and* verbal form. There are so many standards to keep track of, they need to be written down so nothing is forgotten. But websites change rapidly—new technologies are added, old technologies removed, there are new products to promote, and new ideas are generated about how to present information. There are also shifts in your industry and the web more broadly. No document listing every standard can keep pace. Instead, the focus should be on treating the standards document as a living, breathing thing that is updated regularly by everybody involved in the website.

More importantly, though, the standards should be discussed regularly, along with the results those standards bring. By doing so, consistency will be valued by the people who manage your website because the people managing the website see the results that come from being consistent and following standards. The more consistency is valued and everybody involved in updating the website regularly discusses what standards work and why those standards matter, the more likely your website is to be kept consistent. The standards document may or may not be kept completely up to date, but the tacit knowledge of those standards is shared, lessening the impact if somebody leaves. This results in better communication about conventions and standards, ensuring your website looks and behaves more consistently for your visitors.

Technical Considerations

* * *

Mobile-Friendly Websites

Part of developing and designing a new website entails creating different versions of the website to support visitors on different devices. At least three versions are typically needed: one for visitors using desktop or laptop computers, another for visitors using tablets, and the third for visitors using smartphones. These different versions are primarily needed because of the different sizes of the devices' screens. Potentially, you need more than three versions, since there can be larger or smaller smartphones, larger or smaller desktop computer monitors, and larger or smaller tablets. Along with adjusting for screen size, you can design and develop features to better support devices that allow touch-based input.

How you create these various versions (and which method is right to use) is outside the scope of this book, since many of the factors involved are overly technical in nature. But the various methods you can use to create the different versions of each device will influence how consistent your website will be across the different devices. To oversimplify somewhat, you mainly have two options for creating a mobile website: a dedicated version per device or one version modified to work on each device.

The first option for supporting multiple versions of your website is to create an entirely separate website dedicated exclusively to visitors using a certain type of device. This allows you to design a website specifically for people using smartphones, tablets, or desktop computers. With this setup, your website's code is programmed to detect what type of device a visitor is using and then returns the appropriate code, text, and design for that device. This allows you the greatest ability to modify your website to support the needs and expectations of the people visiting from certain devices. This is helpful if visitors' needs and expectations differ by device.

However, because you have entirely different websites with entirely different code and design for each device, this method presents a challenge for consistency. As you develop multiple versions of your website, it's possible these different versions won't contain some or all of the same text, images, functionally, or other features. Even if each version contains identical stuff, all those pieces may not be organized in the same way on each device given the different space available on a visitor's screen.

From a website maintenance and management standpoint, if the text or images are the same across different versions, then any changes to the desktop version of the website will need to be made to the smartphone or tablet versions as well—doubling or tripling your internal workload. Along with these problems, each version of your website will have its own specific standards. With so many variations and standards to remember, something will be forgotten and cross-device inconsistencies will grow.

Websites dedicated to a particular device tend to only be a good option for instances where visitors have unique needs on different devices. In these cases, a dedicated website can actually be a useful way to meet the needs of your smartphone visitors, which are distinct from the needs of your visitors using a tablet or laptop. However, these instances are increasingly rare, since people are likely to have similar needs regardless of the device used.

The next solution that allows for more easily meeting similar needs across various devices is responsive design. This is the more commonly recommended solution for supporting different devices.[13] With a responsive design, your website is transformed to fit on larger or smaller screens. Responsive design leads to more consistency, because this method uses the same code and text for your website across all devices. Instead of a separate website for each type of device, the version of your website shown on smartphones is simply an altered version of the website—fewer characters of text might be shown per line or the images might be smaller, but it's the same text and the same images on all versions.

This makes maintaining a responsive design considerably easier. If an image is updated or text is edited, the change is immediately reflected for people on all types of devices. You wouldn't need to remember to make the same change on multiple versions of the website. Also, this simplifies internal standards since the same text, design, functionality, and features are used everywhere.

The downside to responsive design, though, is that certain alterations can increase the risk of inconsistency across device. For instance, you might hide a paragraph of text from view as you transform your website to fit a smaller screen. Unfortunately, a visitor who saw that paragraph of text when visiting your website from their laptop computer returns later from their smartphone expecting to reread that paragraph. Since you've hidden it from view on smartphones, they'll be unable to do so. The same can happen if a design feature is scaled down to too small a size, effectively hiding it from view, or if some part of the website is pushed lower on the page where few people scroll.

To create the most consistency and offer the most support to people visiting from multiple devices, the focus shouldn't be on hiding things that don't fit or scaling down the size so far it's easy to miss. The better approach is to move items around to new locations on the page—anything important to visitors should be moved higher up on the screen so people can more easily access whatever it is they want to find. If anything, you can consider removing decorative design components. But even these should be removed carefully as these types of components can help people remember your website as they move between devices.

Responsive design doesn't always entail scaling from larger to smaller screens. Instead, the design of your website as it appears on smartphones can be transformed to fit on a much larger screen. When using this mobile-first approach, there is often a strong temptation to add text, features, or functionality to fill the space available as you scale up to larger screens. But, adding to the design as shown on larger screens while leaving these additions off smaller screens results in the same outcome of hiding a something when scaling down.

Whether scaling up or scaling down using responsive design, all versions of your website should contain the same items your visitors most want to see. This way, regardless of the devices used to visit, people are able to get what they want from your website.

• • •

Frontend Code: Tags and CSS

It's easy for the code that defines the headers, tables, images, forms, buttons, font sizes, and the other items on your pages to be styled and used in different ways. This is because the tags controlling all of these items can be written in different ways. Everybody updating your website, even if they aren't a developer, needs clear guidelines on what code to use where to ensure formatting remains the same across multiple pages. The code most responsible for formatting irregularities is the frontend (or client-side) code, which determines how things are presented in a browser. (Backend or server-side code connects the website to the server. It plays a role in consistency, but it's more technical than the scope of this book.)

The frontend code includes specifying the HTML tags that are wrapped around certain types of text. For instance, within your text you have headers or subheaders. In HTML, there are six levels of headers, specified as <h1>, <h2>, <h3>, and so forth. Often in content management systems like WordPress or Drupal, you don't have to edit the HTML directly and instead specify which tag to use around what text through a visual editor. Regardless of how the tags are added, you need to know which type of tags should be used on which type of text in order to keep headers consistent.

Per your website's design guidelines, the page title might be wrapped in an <h1> tag while the subheaders are in an <h2> tag. If, however, somebody updating a page inadvertently wraps some of the subheaders in an <h3> tag, the subheaders on that page will look visually inconsistent with the other subheaders on your website. Depending on the exact design scheme, the <h3> and <h2> tags could look almost identical, limiting the visual discrepancies. Or the tags could be styled to look radically different, especially if the <h3> is intended to be used around a different part of text. This would increase the inconsistency and worsen a visitor's potential confusion.

The same holds true for other types of text and tags wrapped around the text. One commonly inconsistent area involves bolding and italicizing. Some people updating your website's text may prefer to avoid bolding while others might be more liberal with the use of the bold tag. Even if one person is responsible for updating the text, some days he or she may use the bold tag more often. These different choices on what to bold can lead to information on some pages getting more attention because those items are bolded as compared to similar information on another page.

For instance, a visitor might access two different pages on your website in order to compare two different events your organization is promoting. If one of those pages bolds the price of the event, the visitor will likely expect the page about the other event to bold the price as well. If it isn't bolded, the visitor might struggle to locate the price for the event, or the visitor might wonder why one price was bolded—is there something special about that event or that event's price? What to bold and why is a separate conversation, but before you add text to your website, you need to decide what your standards are for using these types of tags.

Technically, one web standard that can help enforce consistency is CSS (Cascading Style Sheets). CSS is a language establishing rules for what various text or other parts of your website should look like. Within CSS, you could specify all text wrapped in an <h1> tag should be 42 pixels and be shown in a dark-red color. This can help keep all of the various tags looking uniform (provided the same tags are used consistently within each page of your website).

Instead of basing CSS rules off existing HTML tags, CSS also lets you define your own classes, where you can establish a certain look for a unique feature. Items assigned the class of "featured-item" could be shown in bold and green. While this offers a powerful way to define unique styles for your website, it's only helpful if everybody updating the website is clear on what custom-defined classes exist within your CSS and where those classes should be used—much like the use of any standard HTML tag.

The more important consideration for CSS is having your designer build CSS rules that are flexible and can be adapted to new types of features or functionality easily. Your website will quickly become inconsistent if you have to constantly create new rules every time you have some new type of text or feature to add. Eventually, some of those new rules will conflict with others. In the worst-case scenario, this can lead to a tangled mess of CSS where each page has its own unique use of colors, sizes, or fonts.

Instead, you want the CSS file to define rules specifying basic colors, shapes, or font sizes that can be applied to different kinds of features and functionality—now and in the future. These basic rules will provide a consistent look for common HTML tags—headers, subheaders, paragraphs, images, forms, buttons, and so on. As new features are added, they can rely on the existing CSS rules even if those rules weren't originally written for the new feature. This will lessen your chances of your website's design spiraling into disarray as you make changes.

Behavioral Considerations

• • •

Expecting Consistency

The people visiting your website have visited many other websites before, with some studies finding people visit at least eighty-nine different websites each month.[14] These visits to other websites have led people to develop expectations about websites in general. Since some of the previous websites visited include those similar to yours, people have also developed expectations about how websites in your industry behave.

If your website deviates too far from these expectations, visitors will experience a state of dissonance. This dissonance will lead to people feeling uncomfortable during their visit, and some people might even consider your website to be untrustworthy or unsafe.[15] Instead, with a highly consistent website, visitors will feel a general sense of calm, trust, and ease.

As new people visit your website, consistency and familiarity helps to form their first impressions. People establish first impressions and decide whether they like or dislike something in as little as one-tenth of a second.[16] Along with this, a person's average transient, or short-term, attention span is about eight seconds.[17] Putting all of this together, people are spending very little time when they first visit your website even noticing what is included on it. Based on this little bit of information, they are deciding if they like your website and think it's worth engaging more deeply. If visitors don't like your website, or see reasons to distrust your website, they will leave in just a matter of seconds. The more consistent your website is with other websites, the more people will feel a sense of familiarity, which will help encourage people to stay, engage, and convert.[18]

If people decide to stay, though, they will develop expectations about how the pages on your website should look and behave. You will confuse visitors if your navigation suddenly moves from one location to another (or disappears entirely) as they move to different pages. Similarly, people may not understand why colors or styles change from one page to the next. These changes prevent people from growing familiar with your website, leading to more dissonance and feelings of distrust. This is one reason people decide to leave your website without converting or engaging.

Consistency becomes more important when you consider most people visiting your website will probably only visit a few times while deciding whether to work with your organization. Within these few visits, you need to help people move toward a conversion point. If people require an excessive number of visits just to learn how to operate your abnormal website, the likelihood of people successfully completing a conversion dwindles. By following global best practices and the standards of your industry, your website becomes intuitive to use—within their very first visit to your website, people will get how to use it.[19]

Websites built for long-term, repeat visitations—such as website applications—have an opportunity to introduce new functionality that may not be so intuitive on an initial visit. Repeat visitors can learn how to use this functionality during multiple visits. Few websites fall into this category as most websites must cater to first-time or infrequent visitors. So the safe approach is to assume your website shouldn't operate abnormally until you have definitive evidence to suggest otherwise.

For websites that do have long-term, repeat visitors, like websites geared toward members or repeat customers, you still want to be careful with any deviations made. Just because people visit regularly doesn't mean they necessarily want to spend their time learning about all the things that make your website unique. For instance, an advanced search form intended for repeat visitors still should have fields and buttons that behave like other fields and buttons the visitor has seen before. Where this advanced search form might deviate is in the complexity of the fields—perhaps the text around the fields is written in jargon only a repeat visitor could understand.

As you add in new features intended for a long-term, repeat visitor, you also want to ensure these unique features behave the same across all pages. This internal consistency across your website will reinforce how the new features work, which will help long-term, repeat visitors grow more familiar with the new features and speed up their learning process. In the example of the advanced search form, if it does use jargon that only repeat visitors would understand, then that same jargon should be used elsewhere on your website to help reinforce what those terms mean.

• • •

Distraction

Consistency doesn't just apply to websites. Opening a door would prove impossible if every doorknob operated different from the last. Driving a car would be difficult (and terrifying) if each street had inconsistent rules for how lanes, stop signs, or traffic signals behaved. Without consistently behaving objects everywhere, life would be challenging, requiring greater focus and mental effort to figure out how every single object we must use behaves. Thanks to most objects behaving consistently with similar objects, life is easier.

This helps because people rarely focus all their energy on a single task. Most people who manage their organization's website want to believe their visitors carefully compare the products or services sold and thoughtfully consider every blog post or resource provided. For any tasks where that much mental energy is applied, consistency would hardly matter as people would be able to quickly detect the differences and understand the various inconsistencies. But people rarely pay that much attention to your website, in large part because most people are distracted during their visit.

The biggest distraction comes from other devices. Most people visiting your website are using another device at the same time: 57 percent of visitors on a smartphone are using another device when browsing the web; 75 percent of visitors on a tablet use another device; 67 percent of people using a desktop or laptop computer are using another device.[20] Of these visitors using multiple devices at the same time, only 22 percent of simultaneous usage is complimentary, meaning 78 percent of multi-device usage is people working on something else on another device while also using your website.[21] This distraction means people will have trouble remembering what they are doing during a visit and will likely make mistakes.[22]

Consistency acts as a memory aid to help these distracted visitors. This is why consistency makes life easier, online and offline. As just one example, consistently behaving doorknobs allow people to open doors without thinking about every step. People can carry on a conversation or check their e-mail on their smartphone while opening a door without a problem. The same is true for your website, where the more familiar the conversion process, the less people will need to consciously think about everything that they are doing when engaging or converting.[23]

If your order form works the same as order forms on other websites, and clearly follows external standards and best practices, a visitor can place an order on their desktop computer while visiting another website or playing a game on their smartphone. But if your order process or contact form asks for a unique piece of information or asks for standard information in an abnormal way, people will have to stop whatever else they are doing in order to successfully convert. This may lead to some people converting incorrectly, as will be discussed in more detail in chapter 4. Other visitors may choose not to convert at all because your website seems too hard or might require too much of their attention.

What to Measure

• • •

Split Tests and Time Comparisons

The hard part of maintaining consistency and following standards is knowing when to be inconsistent—sometimes you have to break the rules. If your website looks like every other website, especially your competitors' websites, you risk not being remembered. But too much deviation and you risk negatively affecting your visitors. To know where and when to break the rules requires purposefully testing inconsistencies to see how they perform—what confuses your visitors, driving them away or what keeps people around, engaging and converting.

The best way to test inconsistencies is with a split test, also known as an A/B test, where some percentage of your visitors sees one version of a page and other visitors see a slightly different version. The differences between versions might be a different layout, color scheme, price points, call to action wording, or tone of the text. With a split test, you measure the impact these differences have on a specific type of conversion or engagement. At the end of the test, you can see if the changes being tested led to a significant improvement.

To get the best return on investment when conducting a split test, you want to test bigger changes, because those have more potential for making an impact on how people engage or convert during a visit.[24] Testing two different shades of dark green or altering just two words out of a thousand words likely won't have much impact, if any at all. These smaller changes also won't do much to help differentiate your website from others.

A big change to test is whether to show pricing information. There may be a concern that if you show pricing, the sticker shock will scare people away or lead to misunderstanding. Or maybe your industry's standard is to not show pricing. After all, if none of your competitors show pricing, does your organization want to take the risk of being different? A split test puts these concerns to the test, allowing you to determine if showing pricing will differentiate you from the competition and lead to more conversions.

During the split test, half of your visitors see a version of your website that includes pricing and the other half see a version excluding this information. After some time of letting people see both versions, you can review the results to compare the outcome for each group of visitors. You might find 10 percent of the people who saw pricing converted, but only 5 percent of those who didn't see pricing converted. From this, you could conclude having pricing information available on the website is worthwhile.

Split tests do have a downside: you need enough people visiting your website in order to have a large enough sample of visitors viewing each version of your test. Let's say your current conversion rate is 1.5 percent, but you want to move it closer to the average e-commerce conversion rate of 2–3 percent.[25] Moving from 1.5 percent to 2 percent is an increase of 33 percent. To have a statistically significant result, you'd need at least 9,200 visitors to see each version of the split test.[26]

For larger websites, getting 9,200 visitors to each version may only require conducting the test over a few days or a couple of weeks, as shown in table 3-1. For a smaller website with only a few visitors per month, the split test could drag on for months before you can get meaningful results. By the time is test is over, the results will probably no longer be useful since so many other things will have changed during that time. As a result, the long duration would mean the split test was hardly worth the time invested.

As an alternative to split tests, though somewhat less reliable, are time-comparison evaluations.[27] With a time-comparison evaluation, you make changes, allowing all visitors to see the change. After changes are made, you compare engagements or conversions from a time period with the changes to a time period without those changes.

Visits/Day	Split Test Duration (Days)
25	772
100	193
500	39
2,500	8
5,000	4

Table 3-1. Split test duration calculations based on daily traffic volume. Calculations assume a current 1.5 percent conversion rate, a 33 percent desired increase, and two test variations shown to all visitors (50 percent per split test variation). Calculated by author using https://vwo.com/ab-split-test-duration/.

As an example, you could test adding pricing to your website, but with a time comparison, every visitor sees pricing. Because this test is an attempt to improve the conversion rate, prior to adding pricing you would collect baseline metrics on your website's current conversion and engagement rates so that you know how people use your website when they don't see pricing. After a few weeks or a month of having pricing information on your website, you would compare the baseline conversion rate to the conversion rate during the period where you showed pricing information. Whichever time period has the better conversion rate is the winning change.

Time-comparison evaluations are more problematic than split tests since seasonality and other external factors play a role. You might collect your baseline metrics in June prepared to make changes and conduct a time-comparison evaluation in July—but are June and July comparable? Maybe most of your customers take a vacation in July, or maybe your competitor releases a new product and a huge advertising campaign in July in the middle of your time-comparison evaluation. As a result, you might experience a decrease in July that has nothing to do with the changes you've made. Split tests account for this by testing changes at the same time allowing all versions tested to be equally affected by these external and seasonal factors.

To mitigate this for time-comparison evaluations, you can look at more months when pulling the baseline metrics. By having six to twelve months of reliable metrics, you can establish monthly averages for any metric you are trying to improve. You might find your average conversion rate is 1.2 percent, give or take 0.3 percent depending on the month. Given these historical metrics, if your conversion rate rises to 1.9 percent after changes are made, you can be reasonably certain those changes led to improvements, especially if the conversion rate continues to stay higher going forward.

These historical metrics also allow you to see what a particular data point was the same time last year. Along with comparing to averages, you can then also compare the rates this May after changes were made to the rates last May in order to account for any seasonality. These historical records can also help you track how external factors—like a competitor releasing a new product or seasonal holidays—have historically affected conversions.

Another concern with time-comparison evaluations is that, unlike a split test, the changes being tested are shown to all visitors. The changes you test can result in decreased conversion or engagement rates. If the reduction is significant, this can have severe negative consequences on your organization until you remove those changes from your website. There is a similar risk with split tests. But with a split test, this decrease would only affect the visitors who saw the changes, which lessens the negative impact.

Whether with split tests or time-comparison evaluations, there is a potential for reducing conversion or engagement rates as a result of changes. This reduction, though, is still valuable information because you've learned what not to do in the future. Because of the risks, you want to be cautious in deciding what to test. Start with smaller changes where the risk is lower. As you learn what works (or doesn't), you can use what you've learned to test larger, riskier changes.

• • •

Cross-Device Expectations

People visiting your website from different devices may expect something different from your website based on the type of device they are using. Knowing these differences will help you figure out how to build and update your website for each device. Some studies have found people are more likely to visit a website from their smartphone to complete a simple task or find local information.[28] Along these same lines, other research shows people visiting from a desktop computer are more likely to be focused on a complicated or involved task, like in-depth research or placing a larger order.[29]

If your visitors' needs do differ per device, then creating and maintaining a dedicated website for each device your visitors use will be beneficial to both your visitors and your organization. For instance, you might have a smartphone-only website that shows how to contact your organization along with other basic information, while you have a different website for people using a desktop computer that shows in-depth details about what your organization does and the products or services offered.

Of course, as more people adopt and grow accustomed to smartphones, the differences between the expectations of visitors on one device or another will start to blur.[30] People on a smartphone may want just as many details as people using a desktop computer. If the needs of your visitors are similar across devices, utilizing a responsive design may make sense as this technique allows you to present the same website to all visitors, just with an altered layout to fit everything onto the visitor's large or small screen.

To know how to support your visitors across all devices, you need to know what devices your visitors use and what your visitors' expectations are when they visit from a certain device. As you begin reviewing which devices people use, the first metric to review to understand visitor expectations is what pages people access when visiting on that device. Are people on a smartphone only interested in your contact page, but rarely if ever look at your product catalog? That might suggest people using their smartphone are less interested in shopping around and, instead, want to talk with somebody who can help answer their question more immediately. As another example, you may find visitors using tablet devices are only interested in looking at your photo gallery and do not access the detailed information about the services your organization delivers. Instead, maybe people on all device types may be just as likely to look at any page of your website, indicating visitor needs have blurred.

Beyond reviewing what pages people are looking at, you can also look at differences in the ways people on different devices engage or convert. You may find people visiting your website from a smartphone are more likely to donate money to your cause, while desktop visitors are more likely to view your webinars or read articles. Or, here again, you may find no differences at all, which is still a very meaningful result to help you learn who your visitors are.

If there is quite a bit of difference in the pages being visited or the amount of people converting or engaging on a given device, then there is a case to be made for having a dedicated website for each device type people use to visit. The dedicated website for a device could promote what's of interest to people visiting from that device. Using a responsive design for people on that device wouldn't satisfy those visitors because these visitors don't want a different layout, they want a different website.

This device-specific website means your organization must maintain multiple websites. However, the differing needs of visitors suggests the burden could prove worthwhile because it presents an opportunity to reach different kinds of visitors, giving them different ways to engage or convert. Before deciding on whether to create a dedicated website for each device your visitors use, you need to ensure the differences are big enough to warrant all the additional work required of your organization to develop and maintain multiple websites. If it's a slight preference for one link over another within the website's navigation, then a small change to the layout via responsive design might be enough to satisfy those unique, device-specific visitor needs—assuming a change needs to be made at all.

Before making any changes and before deciding to create a device-specific website, you also want to check your website on that device to see if the differences result from a technical error. People will be less likely to place orders from their smartphone if order buttons overlap on a smaller screen. People won't visit pages from a tablet if the navigation links are crowded together and difficult to tap. People on a desktop computer may be uncertain how to engage or convert if the text and other features are too spread out across their larger screen. If the differences can't be explained by technical difficulties, then you know you have real differences in what people expect from your website based on the device used. You should evaluate these differences to find the best way to support your visitors.

Supporting Measurements

• • •

Design and Text Inventory

The first step to improving internal consistency is knowing how far your website deviates from standards currently. The easiest way to do this is to take an inventory of what all the various types of items on your website look like—paragraph text, headers, subheaders, buttons, form fields, links, tables, images, calls to action, and more. The easiest approach is to go page by page, taking screenshots of each item. Group these screenshots by item type—all tables, headers, subheaders, and so on grouped by type. Once completed, view all the screenshots for one type of item to see how they compare. Are headers all the same size, font, and color? Are some links underlined while others aren't? Are similar types of images—like all product shots—the same size or shape, and do they follow a consistent theme?

Similar to the inventory of design features, review the words and phrases used to describe the key concepts on your website. For instance, in a travel agency, are the products sold itineraries, vacations, packages, getaways, trips, adventures, journeys, excursions, or something else? Name differences apply to almost every industry as products or services can be called many different things—including internal abbreviations, product codes, industry jargon, popular terms customers use, or a popular brand name. Along with names, check for inconsistencies in how product features, the benefits derived from your services, or pricing information are discussed.

The temptation is to quickly fix any inconsistencies found during this inventory. But before making changes, you need to know which variety works best for your organization and your visitors. For example, if page A has a blue table with small font sizes and page B has a green table with big font sizes, you may initially be inclined to use the green format because you like that better. But as you dig into conversion and engagement rates, you may find people spend more time and are more likely to contact you if they visited page A than if they visited page B. This example would suggest despite what you may prefer, visitors like blue tables and small font sizes. If the metrics aren't clear which inconsistent format is better, this is an opportunity to conduct a split test or time-comparison evaluation to decide what the standard should be.

In some cases, the impact an inconsistency has on your visitors is relatively minor. Some pages may use different fonts or colors for the subheaders within that page's text. This inconsistency could cause some visitors to struggle with skimming your website's text or comparing information across these different pages. But you may find both design formats have strong engagement and conversion rates. This indicates your visitors aren't affected by the inconsistency. You could make these pages more consistent, or you could split test these differences, but doing so likely won't bring much benefit to your visitors or your organization. Instead, the focus should be on improving the bigger inconsistencies that have a measurable impact on engagements and conversions.

Many people stop here and only take an inventory of their own website. It's equally valuable to repeat this exercise on competitors' websites. Take screenshots of every item included in their design and review what terms they use. This will help you get clear on what the external consistencies are within your industry and help you determine how you compare to those industry standards. With competitors, you will not be able to tell how the inconsistent pages or features perform, but anything different about a competitor's website presents an opportunity for something to test, either via split test or time comparison, on your website.

• • •

Inactivity Analysis

Typically, actions are what get measured. What links do people click or tap? What parts of the page do visitors read? What calls to action capture the most attention? In this rush to track actions, there is a key metric missed: What are people not clicking, tapping, reading, purchasing, or submitting? There might be fourteen links people could have clicked on your home page, but maybe only two are being clicked and the rest, including that big call to action button, have no activity at all.

The question is why aren't people interacting with those links, buttons or calls to action, even though they could? There are many reasons why visitors won't interact with a particular part of your website. It could be people aren't interested, or they didn't see how interacting would help them get whatever it is they wanted. But the lack of activity could also be due to inconsistency. Maybe the link they aren't clicking doesn't look like visitors think a link should look. Maybe the majority of visitors didn't understand that the weirdly shaped object on the home page was something they could tap.

An easy way to detect these neglected features is with heatmaps. Heatmaps are discussed in more detail in chapter 2, but the important part for monitoring inconsistency is that heatmaps show a visual report of clicks or taps. With a heatmap, the areas people interact with are highlighted brightly. To identify the areas of inactivity requires looking at the darker areas, where people aren't clicking or tapping. The biggest cause for concern are dark areas around key links, buttons, calls to action, or your website's navigation. No action in these areas directly affects the amount people are converting and engaging.

These neglected parts of a page are an opportunity to find new ways of presenting links or calls to action to get more people interacting. If the area is currently inconsistent, people might be confused by what they would even do with this feature. By making it more consistent and by better following standards, you can reduce this confusion, helping people more easily figure out why and how they should interact with your website.

Alternatively, if the feature people aren't engaging with is very consistent and follows internal and external standards perfectly, that may be an indication that people are skipping this part of your website because it looks like everything else. Because it's so similar, it fades into the background—operating on autopilot, people skim right past this part of your website onto something more interesting. By breaking consistency, either with how the item is designed or how the text is worded, you might be able to get more people looking at and interacting with this currently ignored item.

If after making changes to the item people still aren't interacting, another change to consider is removing the item altogether—after all, nobody is expressing interest or desire for it. No matter what you try with changes to the design or text, people just don't want what that button, call to action, or link has to offer. More than likely, the inclusion of this item is getting in people's way and preventing them from finding something of greater interest. By removing the item, you have the potential to get more people interacting with other parts of your website's pages.

Instead of changing around the existing item or removing it, you could consider replacing it with a different call to action or a different link to a different page altogether. Maybe people could be interested in something on this page, it's just not whatever it is you have currently shown them. For instance, maybe the neglected area gives people a way to email your organization but if you change that to a way for people to connect with you on social media, you may get more people interacting.

Evaluate Your Website

Internal Consistency

- Which design features, functionality, text, images, navigation links, or other items encourage more people to engage and convert while visiting your website? Are these items used consistently throughout the website? If not, why not?

- Is the design visually consistent? Are the header, footer, breadcrumb, and sidebar navigation links located in the same place on every page of your website? Are colors, fonts, and other design aesthetics used consistently across all pages?

- Are all the links to a given page worded similarly? If there are differences, would the words used in the link still make it clear these links take a visitor to the same page?

- For inconsistent links, would visitors still be clear on what type of information they can reach by clicking on the link? Or, do the differences cause visitors to be confused about the link or misunderstand what type of information is contained on your website?

- Is the tone and style of the text consistent across all pages? If there is a change in tone or style, is it intentional—like a page you are testing or a page requiring a specific kind of tone?

External Consistency

- What are the standards for websites representing other organizations in your industry? Where do these websites deviate from global web standards? What makes websites in your industry unique or different?

- What features or pieces of information do websites from competitors or similar organizations in your industry include? If your website doesn't include these features or pieces of information, is the lack of inclusion beneficial or detrimental to your organization and your visitors?

- Are the words and phrases used on your organization's website similar to other websites in your industry? If not, is the break from the industry standard intentional with measurable beneficial results?

- Does your website contain unique text, features, or other items to differentiate it from all other websites? Are those unique and nonstandard items understood by visitors? Or, are these items causing confusion?

Measurement Guide

For guidance setting up the tools for these measurements, see:
http://www.matthewedgar.net/elements/standards

Monthly or Quarterly	Review engagements, conversions, and pages viewed on each type of device visitors use. Are there any technical problems causing differences in how visitors interact per device? How can you promote the things those visitors are interested in? What are the inactive areas of your website where people don't click, tap, or scroll? By making these areas more or less consistent, can you increase activity? If changes don't increase engagements, consider removal or replacement.
Quarterly or Annually	Take an inventory of your website's text and design features to determine how much inconsistency exists. As you identify inconsistencies, measure engagement and conversions to see if the inconsistency is beneficial. Take an inventory of your competitor's text and design features to see how consistent their website is and what changes they have recently made. This will also help you determine the industry standards your visitors expect your website to follow.
Before Major Changes	Before making changes, measure engagement and conversion rates on the page you are about to change. After changes are made, measure engagement and conversion rates again to see if the rates improved. Note, a decrease in engagement or conversion is not a failure. That decrease is a lesson you have learned about what your visitors want and how you should change your website going forward.

120 | Matthew Edgar

CHAPTER FOUR

PREVENT AND HANDLE ERRORS

I N COMPLEX SYSTEMS, things won't always work as intended, and websites are most certainly a complex system. The website is hosted on a server (or servers). Backend code communicates with the server to retrieve the necessary data and files to load each page. Frontend code takes the data and files returned from the backend code and transmits it to a variety of browsers and devices. Browsers have to take all the code received and output it to human visitors. Those human visitors are subject to their Internet service provider's speeds and other restrictions. Once everything has been downloaded into the browser, humans (the most complex component of this system) have to read, interact, and navigate through your website—hopefully without being confused or lost.

Every piece of the process can break—servers can fail, code has bugs, browsers can display code in different ways, each device has its own peculiarities, and humans can be unpredictable in what they will do when visiting. If you think, "My website is so small and simple that nobody will encounter any problems," just take a moment to think about every step of the process as well how much of the process is not in your control. Are you convinced all components will work correctly for every visitor?

No website will work correctly all the time for every visitor regardless of how big or small it is. While work can and should be done to keep the number of problems people encounter to a minimum, things are still going to go wrong. These problems can lead to lower conversion and engagement rates. In some cases, problems can be so devastating that they harm the success of the website and, possibly, the organization running the website. But there are ways to handle the problems that occur. If handled correctly, with the right error messages and support for visitors, you can keep people who encounter a problem engaged and converting.

Key Concepts and Questions

• • •

Resolving Technical Errors

The first type of problem that can occur on websites are technical errors—the server crashes, links break, bugs find their way into your code, or some browsers don't display parts of your website properly. To the extent technical errors can be fixed, they should be fixed as quickly as possible. After all, fixing the error is the most definitive way to ensure nobody reaches that error again.

How to fix these technical errors is well beyond this book's scope, but within this book's scope is sometimes the harder question to answer: How do you decide what errors to fix first? Related, some errors are hard to fix—how do you know if it's worth the investment?

In many organizations, errors are prioritized by difficulty of technical resolution. The programmers might fix simple errors first in the interest of getting the easy stuff off the to-do list. Or they will start with the biggest and hardest errors first often because it's easier to get time and money set aside for those. Some prioritize errors by the opinions and whims of the head person in charge—a HiPPO prioritization. Still others have systems that are more arbitrary, essentially playing a game of whack-a-mole with errors, addressing problems as they arise with no real thought given to priority.

Figuring out what errors to fix first shouldn't be so random or opinion-driven a process. It also shouldn't involve wildly running from one error to the next. As well, figuring out what errors to fix first shouldn't revolve around the technical difficulty of the fix. All of those approaches to prioritizing errors will quickly lead to a backlog with visitors increasingly frustrated by your dysfunctional website.

The better way to decide what to fix first is to look at how the errors affect your visitors and your organization. Based on the level of impact, you can categorize your errors into four main groups.[1]

- Cosmetic: The error causes a little confusion but not enough to reduce conversions or engagement. Most people won't even notice the problem exists. Cosmetic issues aren't really a priority to fix.

- Minor: The error causes some mild frustration, keeping a few people from converting or engaging, but most visitors will push through or not even notice there was something wrong.

- Major: These errors keep most visitors from converting or engaging. People are frustrated by the error and are likely to leave without returning to your website later. Fixing this error is a high priority.

- Catastrophic: The error makes it so nobody can engage or convert with some or all of your website. People will leave, never to return, and they are likely to tell others how awful your website is too. These types of errors can destroy your entire organization.

Say, for example, an error keeps people who visit using a particular browser from converting by completing a registration form on your website. The people using this browser only account for around 10 percent of visitors, but because of the error, none of these visitors are able to convert. This could be categorized as a major problem since people in other browsers are able to convert without issue. However, even though it's a small number of visitors, this example error could also be categorized as catastrophic—especially if the affected visitors are rather vocal in sharing their frustration with your website on social media networks.

As another example, a problem on your server may prevent a sales-inquiry form from delivering messages to your organization—people may think they are converting and sending you their information to you, but nobody at your organization is receiving this information. As this affects every visitor, this is clearly a catastrophic issue and fixing it should be given the highest priority.

Some catastrophic errors can be persistent and affect a large number of visitors for weeks, months, or even years. Fixing these errors can be expensive, and for many organizations can require more time or money to fix than is available. While you want to do what you can to completely fix the error—or at least partially fix—the better answer may be to prevent people from reaching the error-prone part of your website.

The most egregious and persistent issues may require shutting down sections of your website for the duration of the fix. In extreme cases, catastrophic errors require significantly altering your entire organization to focus on something less technically daunting. Returning to example of the broken sales-inquiry form, you might have to remove links to the form until it's repaired. Or, if it can't be repaired, maybe a simpler inquiry form needs to be developed, or visitors need to be directed to call to make a sales inquiry.

As an example of a minor error, visitors on a smartphone may have trouble scrolling through a long informational table. They have to zoom and scroll left-to-right extensively in order to read everything in the table. But people really want to read the information, so the majority of your visitors put up with excessive zooming and scrolling. A healthy percentage even end up converting after reviewing the table. Because people are still engaging and some are even converting, fixing this error isn't a high priority and likely the fix for this error wouldn't be worth the investment of time or money. Of course, if people weren't looking at the information contained on the table and instead chose to leave, this error would move into the major category, requiring an immediate fix.

Cosmetic errors are the last thing to fix. If anything, adding text offering hints about how to work around the issue are usually sufficient. Let's say after people conduct a search on your website they are presented with results and a link to go back to conduct another search. But the link to conduct another search may not work in certain browsers or might be hard to tap on a smartphone. Fixing the link might take a developer a full day given the complex nature of the code, which is a lot of time for such a small error. In just a few minutes, though, you could adjust the text on the search results page to tell people to use their browser's Back button to try another search. The error still exists, but this quick work around will probably be good enough to keep any visitors confused by this cosmetic issue engaged with your website.

• • •

Handling Slips and Mistakes

No matter how technically flawless your website is people visiting your website can still encounter problems during their visit. These aren't technical errors but accidents—clicking the wrong link, forgetting to fill out a field on a form, or deleting an item from a cart the visitor actually intended to purchase. The accidents can be described as either slips or mistakes: slips are purely accidental actions, and mistakes are things a visitor thought they did right but were actually wrong.[2]

The easy answer is to blame the visitor for these slips or mistakes. After all, the website isn't technically broken. The code, server, and browser all worked as intended. If anything, the person's brain is broken. It's easy to wonder what is wrong with that visitor and why he or she can't figure out how to use your website. Some refer to these types of problems disparagingly as PEBKAC (Problem Exists Between Keyboard And Chair) or ID10T errors. So why bother fixing these nontechnical problems since the person visiting your website is at fault?

The short answer is every slip and mistake is the result of something confusing within your website. You want people to engage and convert during their visit, and you are responsible for making it easy for people to do so. People rarely think the slip or mistake is their fault. Instead, visitors will blame these problems on your website, and, indirectly, people may blame your organization for creating such a difficult website to use. This will hardly encourage people to engage with your website, resulting in fewer conversions. Plus, if your organization relies on repeat visitors to the website, people who encountered problems on your website will be less likely to return in the future, resulting in fewer engagements and conversions long term.

Instead of passing the blame, the better question is how do you prevent these slips and mistakes from occurring? The first step is to identify where the accident points exist. What form fields are forgotten? Which fields do people input the wrong type of entry? What links are misleading and sending people to an unexpected page? What pages are people searching for on your website but unable to find? What links are people accidentally clicking? After locating all the areas where slips and mistakes can occur, the next step is altering your website to help people avoid these accidents.

For example, you might find there is a mistake happening on a sales-inquiry form. In response to a question named "Timeline," people are either providing the wrong type of information or are simply entering no information at all into the field. People think they are providing an accurate response by leaving the field blank—not entering information is their way of telling you they don't have an answer. Every time a person submits your form, an error message appears telling them to try the "Timeline" field again because it is required. Technically your code works, but visitors who keep seeing the error message understandably won't think things are working correctly because they don't see why this field should be required.

To help visitors avoid making this mistake, adjustments need to be made to the "Timeline" field. The changes need to make it clearer to a visitor what types of responses are expected. Instead of giving people an empty text box to input their answer, you could give people checkboxes. These checkboxes could offer options ranging from "1 month" to "1 year," including an option for "I don't know."

However, after adding in the checkboxes people might still skip the field altogether or most people will simply check "I don't know." If people keep skipping the field they will continue to get an error message stating the field was answered incorrectly. If most people enter "I don't know" as their response, this is likely unhelpful information to your organization. Either way, these responses indicate people are still confused. But why would they be now that you've added checkboxes? Shouldn't that have made it easier for people to enter information into this field? What else can be done?

Showing specific options for how to answer doesn't always communicate enough about how to respond. The other change to make might be to rename the field from "Timeline" to "Approximately when do you want to complete this project? (Required)." This new name, or label, on the field more clearly communicates what type of response you are seeking. Where "Timeline" could mean many different things and perhaps scare some people off because they don't have a precise answer, this type of question makes it clear you are only seeking an approximate answer. Also, by adding the word "Required" you communicate to your visitors that they need to do something with this field.

Even with these changes, people still might ignore the field simply because they don't see why they should complete it. In these cases, no matter how clearly you explain the field or how specific the answer options may be, people don't understand how it helps them accomplish the task. This is not a mistake or slip, just a misalignment in expectations about how a visitor should use the form.

A common example with contact forms is a field labeled "How did you hear about us?" While this field is of interest to your organization's marketers, it probably isn't relevant to the visitor's intention of contacting your organization. As a result, no matter how clearly you explain what information to input in the field, the field isn't relevant to the visitor. They aren't making a mistake by skipping the field; they don't see the point in providing an answer. In these cases, the better answer is probably to remove the field altogether, or at minimum not make it required.

As an example of a slip, people visiting your website from a smartphone might accidentally tap on the wrong link while browsing through your website. Technically, the links work and go to a specific page on your website. But perhaps the design is such that it's difficult to zoom in so that a person could tap on the links. Or your design might have links too close together making it hard for people to easily tap the desired link. It may also not be an issue with your design at all and instead could be misleading text—given the words used in the link, people mistakenly thought they'd be taken to a different page altogether.

As with the technical issues, these slips and mistakes should be prioritized based on how they impact engagements and conversions. A cosmetic issue that causes mild confusion, like a link that is a little hard to tap on a smartphone, is less important to correct. A link with misleading text might present a major issue because visitors think they'll get something different after clicking or tapping the link—this perceived deception could harm conversions or engagements. A mistake with your contact form that frustrates visitors with confusing error messages might be a catastrophic issue worthy of fixing now before it affects conversions.

• • •

Recovering from Errors, Slips, and Mistakes

Despite all the work you can and should do to resolve and prevent errors, mistakes, and slips, some people will still encounter an error, will accidentally click something, will misunderstand some text, or will leave a form field blank. Websites are too complex to prevent at least something from going wrong during somebody's visit. When people reach an error, slip up, or make a mistake, your website needs to allow people a way to easily recover so they can resume their visit. The last thing you or your visitors want is for the error, slip, or mistake to derail their visit.

The first step to helping people successfully recover is ensuring the visitors understand they have encountered a problem. You don't want to bog the visitor down with a lengthy explanation, but by offering a brief, precise, and polite explanation of what led to this error, you can help the visitor understand the situation.[3] As part of being polite, a helpful recovery message should avoid blaming the visitor for the problem that has occurred, even if it was due to a mistake the visitor made. As well, the text should be human-readable, instead of using the raw, sometimes confusing, technical error message.[4] As one example, people won't easily understand a message reading "Error 503—Service unavailable," but they would understand a message that says, "We apologize, our website is temporarily down for maintenance."

Visually, the error message should be designed to be easily identifiable and easily readable so the visitor can quickly comprehend that an error has occurred and what to do about it.[5] The objective of designing an error message is to get people moving quickly, so the design should be kept as simple and distraction-free as possible. The sooner people realize something has gone wrong, the sooner they can read the text to determine how they should recover.

For errors triggered after a field is left blank or entered incorrectly on a contact form, the error message needs to appear somewhere the visitor will be looking. Two common locations for error message placement are the top or the bottom of the form. The argument for placing the error message below the form is this is nearer the Submit button people just clicked so the visitor may still be looking at this part of the page. However, after submitting a form, the browser sometimes returns a visitor to the top of the page so an error message lower on the page could be missed. Because of this, some websites choose to put the error message above and below the form, though this can sometimes be intimidating for a visitor. The better answer is to pick one or the other and then ensure your code alters the display to take the visitor to the part of the page displaying the error.

After determining where to notify people that errors exist, the next question is to determine where to explain the detail of which fields are in error. On some websites, the details of every field in error are explained within the message shown above or below the form. This summation of all errors can be daunting, especially if many errors have occurred, so a better answer is to explain the details within the field itself.[6] The field or fields in error are highlighted in a different, more noticeable color, and text is placed immediately next to or above the field explaining the details of that field's error—such as an invalid entry or that it was a required field left blank. This approach can help people more easily identify and focus on the parts of the form they need to correct.

Once people know what went wrong, you next need to tell people how to recover. There are three options for how people can recover: leave, go back, or go forward. Leaving is the worst possible outcome. For your organization, somebody leaving means a missed opportunity for him or her to engage or convert. But leaving is also a bad outcome for the visitor as well—people leaving after encountering an error aren't getting what they came to your website looking for and now have to find some other website.

Going back is certainly preferable to having people leave. But going back makes people retrace their steps, causing them to think harder about how to use your website and how to avoid running into the same error again. At best, people who have to go back will be mildly annoyed by this extra work. But if there are too many steps to retrace, people may decide to stop and leave your website instead.

The third option is a successful recovery where people are able to move forward from the error, slip, or mistake to get what they wanted from your website in the first place. With this recovery option, people get what they want from your website and your organization has another satisfied visitor engaging and converting. To help people recover, you can offer suggestions on how to move forward within the error message. This way, people will understand how they recover without retracing steps or leaving.

Let's go back to the example of the field on a form that a visitor left blank. In the error message stating a field is required, you should also offer the suggestion of what type of response is expected or, at minimum, suggest at least some entry must be provided. Or the way forward from the error message may happen off your website. For instance, if the form field people are leaving blank doesn't apply to them, they may need to contact your support staff instead. If that is the case, the error message needs to clearly indicate how to contact your organization and what type of help should be expected.

To find the best path forward, you want to consider what people wanted from your website before this error message got in their way. People reached the error while trying take some other kind of action during their visit, and the error message is preventing them from taking that action. So an appropriate recovery message offers a way to complete the engagement or conversion they were wanting to complete in the first place.

If people wanted to sign up for your newsletter, but the error message suggests downloading a PDF as a way to recover, the recovery will fail. Sure, downloading the PDF is an easy way forward, and the PDF might even include some of the same material as your newsletter, but that isn't what people wanted to do during their visit. They wanted to subscribe, so the error message needs to help them do that. Ultimately, the recovery option needs to be congruent with what the visitor wants. If it isn't, the visitor will be forced to use the Back button or leave.

The final question is when to deliver the error message. The quicker you can detect an error, mistake, or slip the better.[7] If you can show visitors the message indicating a problem has occurred before they are able to make any more mistakes, you'll prevent a lot of frustration. But if you wait until the last possible minute after people have already wasted their time and energy causing so many errors, you'll cause lot of annoyance.

You can often detect early signs of slips or mistakes. For instance, people who are aimlessly clicking through your website but not spending much time on any given page are probably lost and have probably misunderstood some of the links they have clicked. Or, similarly, you may find people are lingering too long on a page that should be quick to visit. To help these visitors, who may be confused, you could offer a chat box to let people talk with your organization so that they find the pages they want.[8]

On a contact form, you could show error messages as people input their response into a given field. This allows people to immediately correct whatever they did wrong before clicking the Submit button. That is preferable to waiting until the person submits the entire form and showing them a long list of errors that now must be addressed at once.[9] If you can catch the error before it becomes a bigger problem and before the next problem occurs, the better your chances of reducing people's frustration and helping them recover so they can keep engaging and converting.

Technical Considerations

• • •

Form Errors and Required Fields

Most websites include some type of form for searching, contacting, ordering, calculating, or something else. Forms can break for a wide variety of technical and nontechnical reasons. A search form may fail technically by not returning any results, even though there are pages on your website that should have been returned for the person's query. Or a search form might return results, but the results aren't relevant to the visitor conducting the search—technically things worked, but things didn't work for the person using your website's search tool.

As a result, when reviewing your website's forms, along with ensuring the forms work technically, you also need to ensure the forms do what your visitors expect. In the example with the search form, you'd want to test to ensure results are returned when you type in a query, but also test to ensure the results returned seem useful and relevant given the search phrases people use. By testing the search form with the terms or phrases people are likely type into your search form, you can identify problematic phrases that return different results than what a visitor would expect.

Another point to consider as you adjust your website's forms is what fields are required. Likely, your organization needs people to provide a response to each field. If a visitor mistakenly enters partial information into an order form, it will be difficult to fulfill that person's order—which is not only bad for your organization, but bad for the visitor as well. By requiring fields, you can help people avoid mistakenly leaving a field blank. This can lead to a more complete submission, which is beneficial to your organization and your visitors.

Along with requiring anything be entered into a field, the code can also check if a response to the field comes in a specific format, ensuring greater accuracy. Maybe your contact form requires a phone number be entered in a specific format. If somebody forgets a digit or doesn't input his or her area code, an error message would help a visitor identify and correct those issues so that he or she can continue successfully completing the contact form.

But what happens when a person from the United Kingdom visits your website based in the United States and inputs a phone number in a different format than the one your code expects? The visitor is correctly entering his or her phone number, so no mistake is being made. The form fails because, technically, the phone number input is incorrect since the correct number the visitor entered didn't follow your code's stated format. When developing your form, the requirement that phone numbers match the United States phone number format likely made sense—after all, your organization is located in the United States and most of your customers might be in the United States too.

This applies to more than phone numbers. Even visitors within the same country may enter in different information than what your code might expect. For instance, if your code limits the street address to only a certain length, people with longer street addresses will be unable to successfully enter this information. To best support all visitors, your website's technical requirements need to allow greater flexibility in the type of data visitors can enter into a form field. By making fields more flexible, you will avoid these error messages for entries that are incorrect-but-actually-correct.

If you do require fields, you need to let your visitors know. The easiest route is to simply add the word "Required" next to the field's label or to use an asterisk indicating the field is required.[10] If you do use an asterisk, best practices state somewhere above or below the form you should indicate what that asterisk means to avoid any confusion. This lets visitors know immediately which fields they need to not miss when completing the form.

The problem with labeling a field as "Required" or placing an asterisk next to every field is that this might result in people skipping the optional fields to focus solely on the fields you said were required—the word "required" is often translated by visitors into "most important." Research suggests you will find more people completing required and optional fields by only indicating which fields are optional.[11] This is because people tend to share more information than necessary by default. By labeling a field as required, you discourage this default behavior and encourage people to only complete what they must.

The next question is deciding how to indicate that something went wrong to a visitor. One common design style involves showing the visitor a long list of every error in their form submission. The visitor can then work through this list to correct each error before resubmitting the form. This seems like a simpler option since all the errors are contained in one location. You are giving a visitor a helpful list of every error they must correct.

However, with this option, a visitor has to continually reference the list of errors as they work to correct each field. As they fix one field, they must scroll back up to the top where the list is located to view the next issue. As well, visitors can start to perceive a long list of errors as one, long task to complete, which is why this can seem so daunting to visitors—especially if there are many errors they have to correct on this form.

The alternative is to use inline feedback messages to display the errors that have occurred. This places the error message next to the field with the error, usually with an indicator around the field that is in error. With inline errors, visitors don't have to refer back to the list of errors at the top of the form and can work their way through each field in a more straightforward process. This makes each error into a singular task to fix one by one, instead of a long list of all errors to be addressed at once. Visitors perceive the inline errors as less daunting, making it more likely they'll correct the problems. Figure 4-1 shows an example of each style.

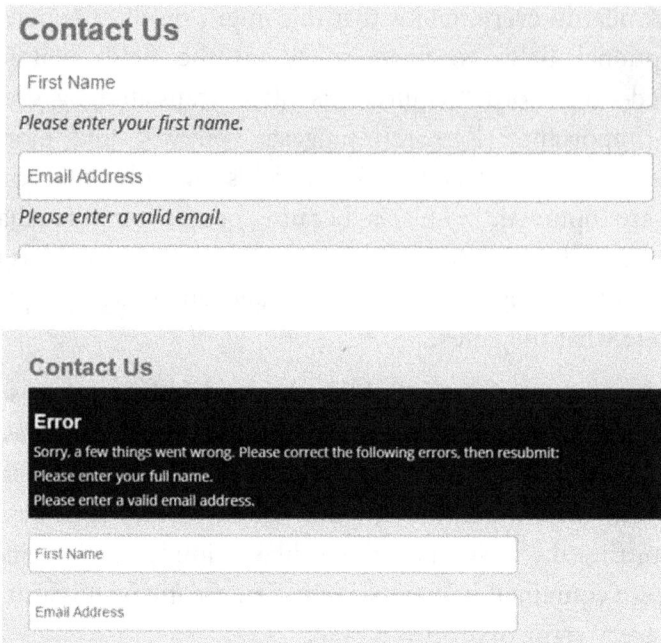

Figure 4-1. Two different styles of form errors. Top: inline errors placing the error next to the field in error. Bottom: a list of all errors that have occurred in the form.

Along with impacts on visitors, there are technical considerations when deciding how to check for errors. To oversimplify somewhat, there are two methods of checking for errors.[12] The first option is to check for errors after a visitor submits the form. Typically, these error checks are run within the backend code on the server. The backend error check code reviews each field against a series of rules—which fields aren't allowed to be blank, which fields are in the wrong format, and so on. After all the fields have been checked, the list of errors is returned to the browser and displayed—either inline or as a long list—for the visitor to review. The biggest problem with backend error checks is that there is always a delay as all field entries must be passed to the server to be run through the code before the errors are passed to the browser.

As an alternative, the error checks can be performed before people submit the form. These immediate checks are handled by frontend code. The frontend code, in this case JavaScript, listens to each field for what's called a blur event, which indicates a visitor moved off the field. As soon as the JavaScript code detects this blur event, the error check can be triggered and, if something is wrong, an error message can be shown. Unlike backend code, there is virtually no delay as all of this process happens in the visitor's browser. Because of the immediacy, people haven't had a chance to move beyond the field yet before the error message appears letting them know something is wrong with that field.

The downside to using frontend code is that some visitors can bypass it by turning JavaScript off in their browsers. As well, JavaScript error checks are dependent on each browser working correctly, which may not always happen. If too many visitors bypass the error checks, either intentionally or due to a browser issue, this will affect the quality and accuracy of your form submissions.

The better technical answer is to use both techniques. Checking for errors as a visitor moves through the form can help people detect and recover from errors quickly. After the visitor completes and submits the form, you can rely on the backend error checks to ensure nothing was missed by the frontend error checks. If something was missed, the backend code can return the additional errors to the visitor. Using these techniques together ensures no errors are missed or bypassed, resulting in more successful and accurate submissions.

• • •

CAPTCHAs: Are You Human?

One common technique used to reduce junk form submissions is the CAPTCHA, which asks visitors to prove that they are human and not an automated program. CAPTCHA is an acronym for Completely Automated Public Turing Test to Tell Computers and

Humans Apart. CAPTCHAs exist because automated applications have been created to scour the web looking for contact forms. The applications can fill out these form and submit spammy information.

Older versions of CAPTCHAs took the form of distorted letters or numbers on a warped background with letters heading in different directions. The automated applications use to be unable to read the numbers or letters, meaning a human had to actually submit the form. But some applications are now smart enough to figure out these letters reducing the effectiveness of these older CAPTCHAs.

Because of the problems with older style CAPTCHAs, Google and others have developed image-based tests, which are more challenging for applications to complete.[13] Other new CAPTCHAs ask people to complete a short puzzle or solve a math equation to prove they are human. Despite these newer options, many websites still use older style CAPTCHAs, with distorted letters, within their forms even though newer alternatives exist.

The spam results submitted by the automated applications can be a mild nuisance if it's only one or two spam submissions every so often. However, spam submissions can become a major problem if so many are submitted that an organization is unable to respond to legitimate submissions. In extreme cases, these automated applications can also submit so much spam through a website that it crashes the website altogether.

The reduction in spam is beneficial to your organization, but using a CAPTCHA has potentially negative impacts on visitors. CAPTCHAs are a common cause for mistakes if people aren't able to read the letters or solve the puzzle. A study from Stanford showed people could only agree on what the letters said 71 percent of the time in older style CAPTCHAs because of how distorted the letters were.[14] Not surprisingly, these old-style CAPTCHAs have been shown to reduce form completion rates, negatively affecting conversions.[15] Even the newer image-based CAPTCHAs have the ability to reduce the amount of people completing forms—in some cases up to a 73 percent decrease in form submissions.[16]

The reason for the decline in form submissions and conversions is because if people make a mistake entering in the CAPTCHA response, they probably won't try to respond to the CAPTCHA again. Instead, they will leave in search of a website that is less difficult to use. Many people are annoyed by the existence of CAPTCHAs and annoyed by the websites that use the CAPTCHA.[17]

Even if we assume no mistakes are made by the people entering in the CAPTCHA response on your website, or if in the future some new version of CAPTCHA exists that is far less annoying to complete, the CAPTCHA still requires a visitor to take an extra moment to prove he or she is human. This slows down the process of completing the form and adds one more barrier your visitors must overcome in order to engage and convert.

This leads to a tough question with no right answers: CAPTCHAs can help limit technical problems and make it easier for your organization to process form submissions without having to contend with lots of spam. But CAPTCHAs cause visitors to make mistakes and take more time completing forms, which will reduce engagements and conversions.

Do you err on the side of the needs of your visitors and remove CAPTCHAs so people don't make mistakes or grow annoyed? This risks spam flooding your mailbox and crashing your server. Or do you err on the side of your organization and technological stability, ensuring your mailbox and website are protected from spammy form submissions, even if it annoys your visitors and reduces conversions?

• • •

Not-Found Errors and Redirects

Another common error, especially on larger or more active websites, is a not-found error—also called a broken link or 404. When a visitor clicks a link to your website, instead of finding the page he or she expected, the visitor sees an error stating the page he or she was seeking could not be located. These error messages frustrate your visitor because, to them, your website has failed to meet his or her expectations in locating the desired page.

The most likely cause of the error is that you have removed a page from your website or moved the page to a new location. Despite the changed location or the removal altogether, people are still able to find links referencing the page at the old location. A seemingly easy fix is to update all the links referencing a page as soon as you remove it or immediately after you move the page to a different location.

Unfortunately though, this seemingly easy fix won't work because many of the links referencing your website's page are out of your control. Search engines return some pages in search results, people may have bookmarked a few pages, other websites link to those pages, people might have shared those pages on their social media profiles, and more. In short, while you can and should update all links that are in your control (such as links on your website, your e-mails, your social network), doing so won't correct all the links referencing a moved or removed page.

To prevent people from accessing a page at an old location or a page that no longer exists, you want to add redirects upon moving or removing a page. Redirects instruct a visitor's browser to send the person from one page to another. It's relatively easy to decide where to redirect a page that has moved to a new location. If your About page moved from about-us.html to about-our-amazing-and-awesome-company.html, a redirect would instruct browsers to take anybody seeking the old URL of about-us.html to the new URL of about-our-amazing-and-awesome-company.html. Anybody finding a link to that old URL would immediately find themselves at the new URL, never encountering an error.

The harder redirects to add are ones for removed pages. The page was probably removed because you no longer want people to see this page—it might have contained outdated information or discussed a product or service your organization no longer provides. But visitors clicking links to the page you removed still expect to find whatever use to be there. Since people can't reach the removed page, you can redirect them to some other page on your website.

The page you redirect to should hopefully be at least somewhat related to the page you removed. If you removed the page because the text was out of date, maybe another page with updated information exists. If you removed one of the products your organization sells, maybe there is another similar product or a newer version people might want to purchase instead. Or if you removed a page discussing an event that occurred in the past, a redirect to a similar event in the future might be of help to a visitor.

You want to do your best to find a page that discusses a similar topic and meets similar types of expectations as the page your visitors were hoping to find. This way you prevent people from seeing an error message and can instead take them somewhere that will hopefully help them find whatever they were seeking. However, if in the interest of preventing people from seeing an error message you redirect to a page that isn't at all similar, visitors won't have their expectations met, and they will be confused or frustrated as to why they have arrived on a page that isn't at all what they were seeking.[18]

When deciding on where to redirect, the home page should largely be avoided in almost every situation. The home page is typically broadly focused, so the chance the home page will be similar to the removed page people were expecting to find is unlikely. Let's say a visitor clicks a link thinking he or she will be taken to a page about red T-shirts, but that page no longer exists, and the visitor is redirected to the website's home page. Because the website's home page broadly discusses all types of clothing this company provides, it won't do much to satisfy the visitor's expectations of finding a red T-shirt. Some visitors may search through the home page to find the red T-shirts they were seeking, but many others will find another website that will make the search for the red T-shirt less challenging.

Instead of redirecting to the home page, you can consider redirecting to a category page. For instance, in the T-shirt example, a redirect to a page discussing all T-shirts available might be at least somewhat more aligned with what a visitor expects to find after clicking the link to view red T-shirts. People redirected to the main T-shirts category page might have an easier time finding the red T-shirts they were seeking, leading more of the people redirected here to engage and possibly convert.

But if no relevant page exists to redirect to, letting people see an error message indicating the page wasn't found can be beneficial—especially if the error message explains why the page the visitor was seeking was removed. In the T-shirt example, if red T-shirts are simply no longer available, it would be a disservice to the visitor to redirect them anywhere else. It would be more helpful if the website simply said, "We no longer carry red T-shirts." While the error obviously won't meet visitor expectations, at least a visitor seeing the error will know why and can move on to find the red T-shirts elsewhere. Thanks to the error message, visitors won't leave confused or frustrated about why your organization redirected them somewhere unhelpful.

Behavioral Considerations

• • •

Friction, Resistance, and Flow

Friction is best explained by MarketingExperiments as a "psychological resistance to a given element in a sales process."[19] Although primarily thought of as having an impact on sales, or conversions, friction also keeps people from engaging.[20] The errors, mistakes, or slips people encounter create friction points, causing resistance. The more resistance people feel toward your website, the greater the error's impact will be for your visitors and the greater affect the error will have on your organization.

As you consider what errors, slips, or mistakes are catastrophic, major, minor, or cosmetic, you can think of the problems in terms of the amount of friction caused. As you fix errors, better handle the messages for the errors you can't fix, get rid of the places people frequently slip, and alleviate mistake-prone items, the less friction people will experience. Resistance gives people an excuse not to engage or convert, and the removal of friction is, in many ways, a process of reducing these excuses.[21] The more you can make changes and fix the bigger points of friction, the less your conversions and engagements will be affected.

Ideally, you want your visitors to experience a state of "graceful flow" as they navigate through your website.[22] In this state of flow, people are able to find the products or services you offer, read your blog posts, contact your organization, make purchases, download resources, donate to causes, and complete other engagement or conversion tasks with no complications. When an error arises or a mistake occurs, this state of flow is disrupted, causing friction. This is why simply stated error messages that offer advice on how to move forward are so important. The more helpful the error message and the more quickly the message is shown, the more easily people can recover and move forward, returning to a state of flow and continuing to do whatever it was they were doing.

An unhelpful error message asking people to go backward makes people retrace their steps. When people move backward to recover from an error, they have to remember what the previous steps leading up to this error were so they don't repeat those steps and encounter the same error twice. If anything, asking people to retrace their steps not only doesn't alleviate friction, it likely adds friction to the process.

As an example of a helpful, friction-free error message, your not-found error message might offer suggestions of other pages the person would like to visit, or possibly offer a search tool to let people find something of greater interest than this error message.[23] Where possible, you can tailor the options offered in the not-found error message. For instance, if many people encountering the not-found error message were looking for pages about products you recently removed from your website, the error message could explain those products are removed but provide links to similar products people may find interesting.

As another example, an error on a contact form can alleviate friction by making it clear if the error is due to the person forgetting to enter something into a field or due to something entered was in the wrong format.[24] If people receive an error on the phone number field, is the error because they didn't enter a phone number at all or because they forgot to enter in their area code? People can probably figure what went wrong with the phone number field on their own, but figuring it out causes friction and disrupts a visitor's state of flow. A clearer error message stating specifically what went wrong and how to correct the error saves people from having to think too hard. This lets them recover more easily and get back to converting by submitting the contact form.

• • •

Slips versus Mistakes

Slips are accidents, happening unknowingly and unconsciously because people are operating on autopilot.[25] Instead of thinking consciously about using every part of your website, habits—developed from using many other websites—guide the visitor along his or her way. For instance, visitors are likely to move through an order form without much thought because they assume your website's order form behaves the same as every other website's order form. If it behaves like everything else, why should a visitor waste brain capacity paying close attention?

However, because visitors are operating on autopilot, they will barely register what your website says. For example, they may not realize your order form defaults to the most expensive shipping option. People slip, submitting the order with that expensive shipping option selected. Only later, when they catch this higher price on a receipt, will they contact you in a panic. Or they accept it this time but will tell their friends how awful your website is at tricking them into such expensive shipping.

One option is to hopefully prevent people from slipping by offering a confirmation page displaying this shipping price a second time. But even the confirmation page can be ignored when a visitor is operating on autopilot. When visitors arrive on the confirmation page, they still aren't thinking consciously—they've seen confirmation pages before and don't see the need to pay close attention. So they click to confirm without ever reading your notice. As a result, they still don't consciously register what the shipping price is or see that it's so high.

A better answer would be to change the default to a lower cost option. This would better support people's unconscious browsing helping people avoid slips, resulting in a more satisfying visit.

Unlike slips, people commit mistakes intentionally and consciously believing they have taken the right steps only to later find out those weren't the right steps at all.[26] Mistakes are due to an inaccurate mental model of how your website works. As an example, on a sign-up form, a field may ask people for a "username." People may think they understand what you mean by "username" and input their first name along with a few digits. Upon submitting, an error appears indicating the "username" should be an e-mail address. How people thought the field worked is not how it actually works.

Some will say people should have known better—after all, next to the "username" field, there was a little question mark icon people could click on to figure out that an e-mail address should have been entered. Why didn't people click the help icon and enter the data correctly?

That's the wrong question because people saw no need to click this icon. People weren't confused by your "username" field (at least initially)—they thought they knew what you expected them to enter. A better solution would be changing the field's label, or name, from "username" to "e-mail address (will be your username)." This change would clarify what type of entry you expect, helping people avoid making a mistake.

What to Measure

• • •

Continuation Rate

For each error, mistake, or slip people are experiencing on your website, you need a way to determine if the problem is catastrophic, major, minor, or cosmetic. The best number to demonstrate how severely visitors and your organization are affected is the continuation rate. Calculating the continuation rate is simple: divide the number of people who recovered from the problem and continued to use your website by the total number of people who encountered the problem. The opposite number is the exit rate, which is the number of people who left after seeing a problem divided by total number of people who encountered the problem. Either way, the continuation rate or the exit rate tells you if people are successfully recovering after encountering the error, making a mistake, or slipping up.

The continuation beyond the error will generally involve people doing something else on your website, like clicking a link to another page from a not-found error or viewing a confirmation message after completing a form. In some instances, though, the continuation point may not exist within your website. For instance, an error message on a registration or order process might suggest people call your support staff to recover and resolve the issue. Measuring how many people take any option offered—on or off your website—gives you an accurate and complete understanding of how well you are currently supporting visitors who encounter errors.

Once you have measured the continuation rate for each problem, you can decide in what order to fix those problems. When the continuation rate is high, the error message presented already sufficiently helps people move forward and recover. This makes these problems minor or cosmetic and a much lower priority to fix. For example, if the majority of people who see an error message about an incorrectly formatted phone number correct the mistake and complete the form, adjusting the phone number field to prevent this error from happening wouldn't be the best use of resources.

The more important problems to fix are where the continuation rate is low and the exit rate is high. Low continuation rates indicate few people are recovering and instead most leave. This means the error prevents people from getting what they wanted from your website and prevents people from engaging or converting. This problem is either major or catastrophic, depending on how low the continuation rate is and how many people are affected.

Along with offering a way to prioritize what errors to fix first, the continuation rate also suggests what error messages to adjust. Fixing the error altogether is an ideal option—especially for catastrophic or major errors—but the fix isn't always realistic given time, budget, or technical constraints. So, when the continuation rate is low and fixing the error isn't feasible, you can improve the continuation rate by changing the message and the recovery options offered.

You may be able to improve the continuation rate by altering the tone of the message. As one example, the continuation rate might increase if the error message is more sympathetic about the visitor's plight and offers an apology along with suggestions for how to move forward. Along with changes to the text, you could also make changes to error message's design. The continuation rate might be low because people aren't noticing the error message at present or the current design makes it challenging to read.

The recovery options included with the error message also influence the continuation rate. Which recovery options to include will change depending on how the people encountering the error expect to recover. The best recovery option, after all, is one offering people a way to move forward toward the thing they wanted before the error interrupted them. As a result, you want to measure—as best as possible—not just the overall continuation rate, but specifically which recovery options people used to continue forward after encountering the error message.

On a not-found error page, people might not recover by using the search box but instead click links to other pages. On an error in a search form, maybe people are recovering by clicking the link to try another search instead of typing in a new search into a search box contained on the page. Maybe nobody recovers at all after seeing the error message on the contact form. Knowing which recovery options people use currently can guide changes to the recovery options offered as you work to increase the continuation rate. These changes might entail adding entirely different recovery options. Alternatively, small changes to how current options are worded could make it clearer how these options would help a visitor recover.

●　●　●

Error Monitoring

It's safe to assume that errors, mistakes, and slips are happening on your website right now—but how do you know which errors, mistakes, or slips? By monitoring for these types of problems, you can identify which ones are affecting your visitors. Monitoring tools can verify if your website is online, detect new not-found errors, see how many people encountered problems with your contact or order forms (assuming such forms are on your website), and see how many people encountered an error with your website's search form (assuming your website has a search form).

The monitoring tools will occasionally be able to help you calculate the continuation rate by telling you how many people are affected by the error and how many people continued beyond the error. But for some errors, these numbers can be difficult to locate (if they exist at all). For instance, there are many monitoring tools to identify not-found errors, but most of those tools don't tell you how many of your visitors are affected by the error, let alone how many continue beyond the error. In these cases, additional tracking through a web analytics tool would need to be added to see how many people saw these errors and continued beyond it.

Along with monitoring for technical errors, you also need to monitor for slips and mistakes. Some mistakes are easier to spot—people may have misspelled a word in a search form, which resulted in people not finding the results they were seeking. By reviewing the words people use to search, you can identify typos and adjust your website's search tool to return results for those common typos.

Other slips and mistakes can mimic normal visitor behavior. How do you distinguish between an accidental and purposeful tap on a link by a visitor on a smartphone? If people delete something from the shopping cart, how do you know if that action was intentional? What is the difference between a visitor who is deeply engaged and a visitor who is horribly confused—both will spend a lot of time clicking around to many pages?

One of the better ways to distinguish between normal behavior and slips or mistakes is to watch recordings of visitors. By watching a recording, you can see if the people who tap on a particular link quickly hit the Back button, which would indicate the tap was accidental. If people delete something from the cart, the recording will show if they re-add that same product within the next few minutes, suggesting the deletion was a slip. If people visit many pages, but don't spend enough time to read the text or watch the videos on the page, this suggests people have made a mistake and lost their way. If you see people scrolling quickly through each page—too quickly to engage—before clicking on the next link, people might be looking for something but aren't sure how to find it.

As you monitor for errors, slips, and mistakes, it's tempting to fix every problem you identify because you want your website to work perfectly. This temptation is why so many organizations struggle with fixing the problems on their websites—everything seems like top priority when you see it in a recording or get an alert from a monitoring tool. Before you review the problems that are happening, remember you'll never keep all of your visitors from encountering some kind of issue. This is why you want to prioritize fixing problems by the continuation rate—by doing so, you'll more easily pinpoint the catastrophic and major problems allowing you to make the changes necessary to help your visitors avoid the most egregious errors, slips, and mistakes.

• • •

Retrace Error Steps

As part of preventing errors and helping people recover when they make mistakes or slips, it's critical to know what caused the problem. By retracing a visitor's steps, you can figure out what went wrong and, from there, make the necessary adjustments to the design, text, or other features to help people avoid the problem in the future.

In some cases, the steps people took to reach the error may seem obvious. If people get an error message stating certain required fields were left blank on a contact form, the cause is people left the fields blank. What more is there to figure out?

This explanation is incomplete because it doesn't explain why people left the field blank. Was it left blank because the text around the field was worded in a confusing way? Was it left blank because of a design issue that hid the field from view on certain browsers or devices? Did people think the field didn't apply to them? Or did they not see the field? The way to prevent this error will change based on how those questions are answered.

Knowing the specific reasons why a field on a form was left blank is tricky. One option is to interview visitors for their thoughts on why they might leave this type of field on blank. Although interviews could clarify the reasons, arranging for and conducting interviews can prove time-consuming, which may not be a worthwhile endeavor to understand one blank field.

The other problem with relying on interviews is that an interview will make people think consciously about how to use the form you are investigating. Because interviewees are paying close attention to the form, they may not miss the field at all. If your visitors leave the field blank due to a slip, this could be because they are completing the form on autopilot. An interview where people think consciously and deliberately about a form won't help you reveal the true cause of why people leave a field on this form blank. Similarly, if your visitors leave the field blank because the words are confusing, interviewees may not experience this same level of confusion because they will probably apply more mental effort than your visitors.

In absence of talking directly to your visitors, you can review your form and make a few educated guesses as to why people might leave a field blank—is it to do with the design, the wording, the positioning, or a misalignment with visitor expectations? An educated guess on why the field is being left blank can lead to quick changes. Following the changes, you can see if there is a reduction in the amount of people leaving the field blank. Of course, the other approach is to ask if you really need that field—sometimes, the best change is removing the field your visitors refuse to answer.

For other problems, the question is less about why and more about how the problem occurs. In the case of a not-found error, people probably reached the error because they clicked a broken link, so the question becomes how people found that broken link. The broken link that led people to the not-found error can come from a wide variety of sources—from Google search results, links on other websites, or outdated bookmarks. To know how people arrived at your not-found error and to fully answer the question of how the error occurred, you need to locate every source. This will also help you identify all of the not-found errors you might need to fix.

Evaluate Your Website

Forms

- Are required fields marked as such in a simple, self-explanatory manner, either with the word "Required" or with an asterisk? Alternatively, have you marked optional fields instead of marking required fields?

- Are error messages written to obviously, clearly, and specifically state the problem that has occurred and how to recover from it?

- Are error messages shown as people complete each field? Or are error messages shown only after the form is submitted?

Not-Found Errors

- Is the not-found error page branded to your organization so that people at least know they've arrived on the correct website, if not the correct page?

- Are links of potential interest or a search form provided to help people who reach this type of error find what they are looking for so that they can move forward?

- Is your organization's contact information provided to help people who need immediate assistance?

- When adding redirects, is the error redirected to a relevant page instead of somewhere general (like the home page)?

General Error Considerations

- Is the error message stated simply, clearly explaining what has occurred and what to do about it without blaming the visitor?

- On smartphones, are the links people need to tap big enough, or could people accidentally tap another link?

- Are there options provided for how to recover from the error? Do the options offer a way forward or backward? Of the options provided, which ones remove friction?

- If there aren't any recovery options on an error, what options would make sense to add—how do people want to move forward after seeing this error?

Measurement Guide

For guidance setting up the tools for these measurements, see:
http://www.matthewedgar.net/elements/errors

Baseline Setup	Set up tracking to monitor the technical errors that could occur on your website—including server outages, not-found errors, errors on all types of forms, or errors within an order.
	Set up services to record visitor behavior on website sections that are critical to engagement or conversion.
Monthly or Quarterly	Review error monitoring services to see what errors occurred, how many visitors were affected, how many visitors continued beyond the error, and what options people used to continue. Errors with low continuation rates are catastrophic or major, requiring immediate attention.
	Watch a sampling of the recordings to identify slips and mistakes that can be corrected either as part of routine changes or large-scale projects. Use these recordings to determine how many people make that mistake or slip up in certain parts of your website.
Before Major Changes	As you fix errors and prevent accidents, review the steps people took on your website before finding the error. Ask yourself why and how the error, mistake, or slip occurred.
	For not-found errors, locate the sources that led people to each error, such as broken links on another website.
	For other problems—where the reasons for the error, slip, or mistake are less clear—develop a few educated guesses of possible reasons why and then make quick changes and watch how those changes affect the number of people encountering the problem.

CHAPTER FIVE

REAL WORLD, REAL PEOPLE

THE VISITORS WHO come to your website are real people with real problems to solve and questions to answer. Whether the problem is finding their next travel destination, looking up an address for your store, or buying a new pair of pants, the people visiting have expectations of how your website will help them. These expectations span how your website is built technically, how your website is organized, and the language used in your text.

The biggest mistake made is building the website to meet the needs of your organization while ignoring the needs of visitors. Neglecting visitor needs is rarely done maliciously or intentionally, but out of the assumption that your visitors share your expectations and think in the same way you do about your website. This is rarely, if ever, the case as your visitors don't share your same level of knowledge about your organization or your website. Or, to put it another way, Arnold Lund's First Maxim of Usability is "Know thy user, and *you* are not thy user."[1]

Instead, as a first step toward matching a user's, or in the case of a website, a visitor's expectations, you need to figure out what people expect. From there, you can design, develop, and write your website to better align what is offered to what visitors want. Complicating matters, each person's expectations will differ, at least somewhat.

This is because each person visiting is unique with varied expectations and past experiences interacting with websites like yours. This wide variety of expectations and experiences can make it challenging to meet everybody's individual needs. But increasing engagement and conversion rates involves accepting that challenge and meeting as many of those expectations as possible—largely by trying to find the commonalties among what all of your visitors need, want, and expect from your website.

Key Concepts and Questions

• • •

Words, Phrases, and Icons

One of the largest expectations people have is the websites they visit will use the same language they do. Undoubtedly, you have certain words, phrases, and jargon to describe what your organization does. It's tempting to use this language within the text of your website. After all, the words you use are technically correct. You are the expert and know the proper way to describe the products, services, and concepts your organization works with every day. It's natural to think other people must also understand these words and phrases, or at least think people will rely on your organization to show them the right words or phrases to use.

However, the people visiting your website, especially those who might be unfamiliar with your organization or industry, have different ways of thinking about what you do. They will use different words and phrases when discussing your products and services—possibly even technically incorrect words and phrases. Even people familiar with your industry might have different ways of thinking and talking about similar concepts than you do. Whether for expert or novice visitors, you can't assume you know the right words to use.

These different thought patterns will shift how you have to explain what your organization does to every person who visits. Since you can't write unique text for each individual visitor, you need to use the words or phrases understood by the majority of visitors. Say, as an example, you are writing a new page that describes an advanced concept related to your organization's products or services. Undoubtedly, you have specific shorthand ways of thinking about and describing this concept. But in creating the page about this topic, you need to select language that matches how the majority of your visitors discuss and think about the concept—even if that doesn't match your internal thought patterns. By doing so, your website's text will be more accessible, usable, and understandable for visitors.

A common recommendation is to avoid jargon or expert-oriented terms altogether. But if your website caters to expert visitors who are highly familiar with your industry, and you know for certain the majority of your visitors are comfortable using that jargon, acronyms, industry slang, or other expert-oriented terms, then using anything other than those terms could be off-putting and even insulting to your visitors. That is, if your website targets senior orthopedic surgeons, your website's text can and should use the language natural to experienced practitioners of that field. Use anything else and those surgeons wouldn't see much reason to visit your website. So instead of assuming you shouldn't use advanced words or concepts, start by understanding how the majority of your visitors talk and think about whatever it is you do, then write your website to match.

Some websites targeting an inexperienced audience might still want to use expert-oriented terminology, even if their visitors don't use these terms. The thinking is this provides a way to teach people more about their industry. If you are considering this approach, remember most people expect to visit your website, get the information they need or purchase the products or services they want, and then get on with their lives. Normally, visitors aren't going to be interested in learning about your organization or your industry, so be careful trying to encourage people to do so.[2]

As you review what words and phrases the majority of your visitors use, you also want to consider what grade level of writing is appropriate. A website targeting a younger audience that is still in elementary school would need to write to a lower grade level to make the text more accessible. For highly educated audiences, you may be able to write at a higher-grade level and still have your text be understandable. For the average website, though, it's often best to write at a seventh or ninth grade level—similar to mainstream novels—as this allows you to target a much wider audience.[3]

Along with the experiences and education levels, the words people expect to see on a website are based on the words they use offline. Those offline experiences form conventions of what things are called. For instance, places online where physical goods are sold is known as an "online store," which compares it to a physical store even though there is no technical reason it needs to be that way. To continue this example, within the online store, items to be purchased are placed in a cart, even though a physical cart is considerably different than a virtual cart.

The word "cart" is used because most people have been to a physical store and put items they want to purchase into a physical cart. This physical object is associated with the virtual version, even if the virtual version behaves somewhat differently than the physical version. As an example of one difference, you wouldn't abandon a full cart in the middle of the grocery store in the same way you might abandon a full cart in the middle of a visit to a website. Despite the differences, there is enough similarity between the concept of cart offline to the concept of cart online that using the same term lets people better understand how to use your website.

Along with words, real-world concepts can also be communicated with icons. A website's shopping cart usually includes an icon of a cart. Some websites use an icon of a house in their navigation to indicate a link to the website's home page, even though a home page is hardly the same thing as an actual, physical home. An envelope icon is often used to represent an e-mail addresses, although no envelope is required to send an e-mail. A phone icon is often used to identify phone numbers, even though very few phones today resemble the 1930s-era phone the icon is based on.[4]

The examples are plentiful, but the point is using icons of physical concepts, even somewhat antiquated ones that vary in behavior from their physical counterpart, can help you communicate with visitors. With very few exceptions, icons work best to support the text as few icons can communicate an entire concept on their own.[5] For instance, depending on the website, an envelope icon by itself might mean people can click the icon to view a page listing all types of contact information. Or, maybe clicking the icon would let people view your street address so that they can send a physical letter. Clicking the icon might instead let people access your e-mail address or open a contact form. A few examples are provided in table 5-1.

Instead of using an icon by itself, placing the icon next to text can help reinforce what that text has said—or, for people who only skimmed the text, the icon can help people fill in the gaps.[6] Back to the envelope example, using that envelope icon next to an e-mail address might help people quickly skim through your website and locate the e-mail address.

The more these words and icons are based on real-world concepts, the more likely it is that people will be able to understand your website. The more people understand your website, the more likely it is they will engage or convert. So, if your website isn't currently engaging visitors or generating conversions, it could be worth rewriting the text at a lower grade level, using different words or phrases, using different icons (or using some icons to reinforce the text if none are used already), or getting rid of jargon (or, depending on your visitors, adding in jargon) to see how if that positively affects engagements and conversions.

Icon	Different Potential Meanings
	Comments, Discussion, Blog, Chat, or Contact Technical Support
	Email Address, Physical Mailing Address, or Contact Form
	Secure Area, Login Area, Password Retrieval, or Checkout

Table 5-1. Example icons and a few different things each could mean when used on a website. Icon credits: Font Awesome.

• • •

Navigation and Call to Action Expectations

Making your website think how visitors think isn't just about the language used. It also includes your website's organizational structure and making sure everything on your website matches how visitors think it should be organized. The organizational structure is most prominently communicated to visitors through your website's navigation. While navigation often refers to a bar or row of links near the top of the screen, a broader definition of navigation would also include any groups of links in the footer or sidebar, along with any links within the text of the page itself. All of these types of navigation together tell people how your website is organized. The more all of the navigation's organizational structure aligns with what people expect, the more people will engage and convert.

One common reason navigation doesn't meet people's expectations is that it's structured the same way the organization behind the website is structured. This style of navigation is often dubbed "org-chart navigation" because it often highlights each department or division as well key personnel or important projects in their navigation. An org-chart navigation may instead reflect how services and products are grouped internally, such as by manufacturer or by project type.

Likely, the internal structure of your organization makes sense operationally given the way internal processes work. This can be an effective way to structure internal resources, including intranets built for your staff, contractors, or other business partners. If you are developing an internal website, utilizing this style of navigation might be appropriate as visitors to that internal website will usually think in terms of your internal organizational structure.

But on a public-facing website that serves a wider audience, navigation based on an internal structure will rarely connect with visitors. Similar to deciding what language to use, your expert view will be technically correct, but most of the people visiting your website will not know (or care) how your organization is structured internally. Visitors have their own ideas about how your website should be structured based on the things they expect to find on your website during a visit. To avoid confusing visitors, your navigation should be structured more in line with those expectations.

As a way to avoid navigation based on internal structures, some websites organize their navigation around different types of people, or audiences, who visit. A website might target doctors and patients, so the navigation could have an area for doctors and a separate area for patients. In a way, dividing the navigation by types of audience seems smart. The text, features, or functionality in each section can be targeted to people in a specific group. Different audiences might use different words or phrases, recognize different types of icons, have different levels of familiarity with your industry, and may even have different levels of education. These different groups might even want to engage or convert in different ways. Certainly, it seems this type of navigation would be superior to org-chart navigation since it's based around people and not around the organization itself.

Except audience-based navigation structures are just as likely to confuse visitors. For starters, people are not always sure which group they belong to and, therefore, won't know where to click. Part of this confusion is because the internal terms your organization uses to describe the different groups of visitors may not match how people would describe themselves.[7]

Let's say a nonprofit in the finance space serves financial professionals as well as consumers looking for information about financial services. This example organization has a Business section containing information for financial professionals, and it's called Business because these visitors use the website in a business context. They also have a Consumer section for nonfinancial professionals. This seems better than an org-chart based navigation with links to various types of loans, securities, legal considerations, or trader information. But what if a visitor is not a financial expert and the information needed concerns his or her business? Most people would probably assume they should click Business even though they should choose Consumer.

Another problem is audience-based navigation styles also run the risk of having to duplicate pages because multiple groups of people might need to see some of the same information. Returning to the example of a nonprofit serving financial professionals, what if information about certain types of loans needs to be presented to both business and consumer visitors? Should a link to the same page be included in each section? That might lead to confusion on which link to click. If it's the same page linked to from both sections, can the page be written to serve the expectations of both groups? Financial professionals and consumers may want different things from the page and have different expectations for what words or phrases will be used to describe the loan.

Alternatively, maybe two different pages need to be written—one for the Business section and one for the Consumer section. This would let this nonprofit meet the needs of business visitors separate from consumer visitors. However, the pages might be very similar, containing almost exactly the same information, just worded slightly different. This near duplication might confuse visitors who happen to visit both pages. Also, having two pages with such similar information creates a maintenance chore for this nonprofit—any time there is an update to the information requiring a rewrite, two pages now have to be updated.

Instead of duplicating the page or linking to the same page from both sections, should that page be placed in just one section and not even referenced in the other? Doing so would hide it from people looking at one section, eliminating duplication and more clearly specifying which group should access this page. But people might wonder what they are missing in the other section, especially if they cannot find what they wanted within their designated section.

All of these problems stem out of the biggest problem with

organizing navigation based on different audience types: this style of navigation isn't structured the way people think. Grouping your visitors is a helpful internal mechanism to understand and track people—when you make decisions about what to change on your website, it's easier to measure what business visitors do compared to what consumer visitors do. When visiting a website, though, people don't think of themselves as part of a particular group. Each person visiting your website believes he or she is a unique individual with unique needs that your website needs to satisfy.

Even if there is no way visitors could confuse which group they belong to, there is still another problem with audience-based navigation. When people arrive on your website and are asked to select a certain audience type, they have to stop what they were doing to figure out what group they belong to before clicking the link to the (hopefully) correct group.

At best, this breaks up the flow of their visit, slowing people down. It also means people may make a mistake and end up in the wrong group, seeing pages that weren't intended for them—making their visit more complex and confusing than it ought to be. This means audience-based navigation structures reduce the odds visitors will engage or convert.

Another approach is to organize your website and structure your navigation by task. People come to your website for a reason, whether to access information, purchase a product, request a service, download a form, and on the list goes. By showing tasks people want to complete in the navigation, like "Contact Us" or "Sign Up," you can assist people in quickly doing whatever it is they came to your website to do.

This style of navigation gets closer to thinking like visitors think. But people visiting are not always ready to complete a task when they first arrive. They might just want to browse around before contacting you, signing up, donating, downloading, subscribing, viewing a demo, or completing some other task. As well, there are many tasks people could complete during their visit—you can't always show them all. If there are many different tasks people could complete, they may need more time to look at your website and decide which task is right for them.

Another navigation style that gets closer to how people think is topic-based navigation. Upon arrival, people are often thinking in terms of broader concepts or topics.[8] For instance, people might think, "I need shoes" when arriving on a website. They aren't thinking "Buy Shoes" (a task) or "Shoe Buyers" (a group) or "Footwear" (an internal department). So, listing the topic of "Shoes" in the navigation would more closely mirror people's thought patterns. Topic-based navigation is also broad enough to attract different kinds of people with different kinds of expectations to similar sections of your website. A section about the topic of "Shoes" will attract people who are thinking about buying a pair of shoes as well as attract people who simply want to browse or search the shoes you have available.

After people click on the broader topic, they are more likely to next think about the tasks related to that topic. In the "Shoes" example, those tasks might include making a purchase, comparing products, browsing a catalog, or watching videos explaining why your shoes are the best to ever be made. Because people think of topics first, followed by tasks, organizing your navigation by topic-then-task will usually more closely match how visitors think.

The next question to consider is how to tell people which tasks they can complete after selecting a topic. Typically, once people have navigated into a topic, calls to action become the primary means of people navigating from one page to the next. As well, calls to action can serve as a guide leading people down the path toward an engagement or conversion. The problem with calls to action, though, is if they aren't relevant to what visitors expect. Chapter 2 discussed how calls to action can become overwhelming if messages take away too much of a visitor's control and chapter 1 discussed the problems of making a call to action too complex. These problems occur because of a lack of relevancy.[9]

The problem isn't so much the pushy tactics or a particular design, but rather the website is coercing people into doing something they don't want to do. Maybe people want to read your blog posts about a particular topic (an engagement), but they aren't interested in subscribing to a newsletter about that topic (a conversion). Or people may just want to research a product but not purchase it (yet, anyway).

Another way of saying this is that the people visiting a website don't really care about the economic needs or your organization. Instead, they are far more concerned with meeting their own needs by getting whatever it is they want to get. Think of it this way: the last time you purchased a product, did you make that purchase to help that company meet their sales goals or because you needed or wanted to buy that particular product?

The tasks you decide to promote within calls to action, like the rest of the navigation, need to be relevant to what people are interested in getting from your website.[10] The best calls to action promoting a conversion convincingly and completely shows how visitors can answer their questions and solve their problems if only they give your organization their e-mail address, phone number, or credit card. Using relevant calls to action to help people navigate through your website toward a conversion will allow your visitors to leave satisfied with their expectations met while simultaneously generating more conversions for your organization.

Technical Considerations

• • •

Response Times

If, in the offline world, you turn a handle and then push (or pull) on an unlocked door, the door will immediately respond by opening, and you can walk through the door to wherever it leads. There is clear cause and effect. When you click a link on a website, you generally assume the link will take you to some other page or will alter the display somehow. But, defying expectations, clicking the link may cause nothing at all to happen—or seemingly nothing. Online, there isn't always a clear link between cause and effect. While the delay could be due to an error, more likely it's due to your website taking too long to respond.

The stuff on the other side of the door you push or pull open doesn't need to load—it's just there. Online, the stuff on the other side of the link needs time to load—the request for that stuff begins when the link is clicked. It's this delay between clicking the link and seeing the results of the click that breaks the cause and effect for your visitors. This is why long response times between clicking a link and the page loading is such a problem. This disconnect between cause and effect leads to impatience over why clicking a link didn't seem to do anything.

How quickly things respond is a primary factor people will use to determine how satisfied they are, or aren't, with a visit to your website—the more satisfied a visitor is, the more he or she will engage and convert.[11] To keep satisfaction high and impatience low, you need to establish a clear and immediate connection between cause and effect by making your website load as quickly as possible to any actions visitors take during their visit. The question is how fast do people expect your website to load?

Google recommends the part of the website people see before they scroll should load within one second on mobile devices.[12] Others suggest the entire website should load in at least two seconds on any device.[13] Some studies show even a one-second delay can reduce conversion rates by up to 7 percent—people would rather leave than wait even one more second for your website to respond.[14] In reality, though, your website needs to load as fast as it possibly can and certainly faster than any competitor websites.[15] There is no such thing as too fast since visitors expect your website to, ideally, match the immediate cause and effect experienced offline, in the nonvirtual world. This means you need to continually improve your website's response times.

The biggest contributing factor to how long it takes a website to load is the frontend of the website.[16] This frontend includes HTML code telling the browser how to present text, images, videos, design features, and the other items on your website. The text loads relatively fast, but images or videos are typically larger in size, which can slow down the load of entire your website.

The frontend can also include JavaScript files specifying different types of interactive functionality or features. It can also include CSS files specifying various rules for how the website looks. While JavaScript, CSS, images, videos, and other interactive items can improve your website's experience—and make it look far better than a fast-loading text-only website—each additional file is one more thing the browser must load before a visitor can start engaging or converting.

Each individual image, video, JavaScript, or CSS file might be relatively small and quick to load, but even lots of small files can add up to large file sizes. If there are twelve images, two JavaScript files, and two CSS files on a given page, there will need to be at least seventeen requests made of the website's server: one for each file plus another request to access the page itself. That example is optimistic on requests required as an average web page includes twenty-three JavaScript files, seven CSS files, and nearly sixty images.[17] This represents a 184 percent increase in a page's size between 2011 and 2016—and this increase in size will likely continue.

Of course, a lot of stuff is required to make your website worth visiting. You need CSS files to create an interesting, memorable design. You need JavaScript files to create the functionality your visitors expect from a website like yours. You need images or videos to more simply explain the products or services your organization offers and to complement the text you have presented. But perhaps there are ways to make your website worth visiting without adding too many additional resources that slow your website's load time.

As you review the pages on your website, you can use speed test tools to identify all the various resources required on your website. As shown in figure 5-1, these tools will also tell you how long each resource takes to load. As you review each of these resources, consider what to change. For example, is that interesting, image-heavy feature on your home page really worth the extra half second it takes to load? Do people need the cool JavaScript calculator in order for their expectations to be met? Or will people engage with a faster but slightly simpler design? Even if a simpler design meets people's expectations, does that simpler design portray your organization and your brand in a way you intend?

To help reduce load time without reducing features or functionality, JavaScript and CSS files can be consolidated into a single document. While it requires extra work as part of development and presents technical challenges that may not always make it feasible, twenty-two JavaScript files could be combined into a single file. That way instead of twenty-two requests being made to the server for each individual file, only one request will need to be made. The same is true for CSS files.

Another potential way to make websites load faster is to utilize some form of website caching. Caching works by creating a temporary storage for certain requests. For example, in a content

Waterfall View

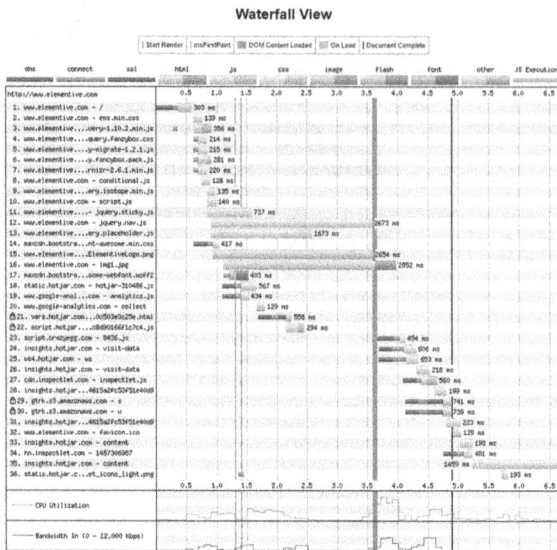

Figure 5-1. You can measure a website's load time using a speed test tool like Web Page Test (http://www.webpagetest.org). The Waterfall View shows you every item required to load a page. Items with longer horizontal bars take more time to load.

management system like WordPress, the text is stored in a series of database tables. Without caching, every time a visitor attempts to load a page, multiple requests need to be made to the database to return the contents of the page. With caching, all of the data requested about the page is saved in a temporary location—this is usually a static text file or another location in the database.[18] However it's saved, instead of several requests being made of the server for every visitor requesting the page, only one file is requested, resulting in faster retrieval.

Along with caching files on the server to limit database requests, you can also use browser caching. When visitors load the first page of their visit to your website, they have to download every CSS, JavaScript, image, and other files to their browsers. If browser caching is enabled, those files are stored within the person's browser. If the visitor clicks to another page, or if the visitor returns to the website sometime later, the files will be loaded from the browser's cache, instead of requesting files from the server.[19]

Despite the many ways available to increase how quickly your website loads and responds to a visitor's actions, sometimes you just can't get a part of your website to go any faster. Maybe it isn't technically possible. Or maybe the cuts required to increase the speed require too many sacrifices to what your website offers. In these instances, you want to increase the perception of speed and find ways to establish a link between a cause (a visitor's actions) and effect (the response to those actions). Remember, a delay of more than a few seconds can make visitors leave instead of waiting for a response.

One common approach is to use loading indicators or progress bars to show your website is responding to the visitor's actions. Ideally, these progress bars do more than show a spinning circle, but instead state specifically how much longer your website will make the visitor wait—either by showing a percentage or a timer.[20] Along with helping to link cause and effect, the use of progress bars can also alter a visitor's perception of time. One study found the use of a backward moving, decelerating progress bar made the process appear 11 percent faster than it actually was.[21]

• • •

Automated Visitors

Not every visitor is a real person. Automated programs, also known as robots, are visiting your website too. The most popular example are search engine robots, like those from Google or Bing.[22] As a brief, simplified overview, search engine robots crawl through as much as the web as they can find and collect as much information as possible about what each page found contains. Then all the information collected is processed and used to determine what websites to show to the people who are searching the web.

Along with search engine robots, social networks also have programs that crawl through a website gathering information. Social networks typically crawl through a website to determine what text or images should be included if somebody shares a link from that website on their network.[23] Many other software programs crawl through the web, such as those creating rating lists or compiling sets of data about different types of web pages.

Many of these robotic programs are intended to help humans determine what websites are worth visiting—especially robots from search engines and social media networks. In older days of the web, the needs of robots and humans differed greatly. Now, however, the expectations of these robots increasingly mirror the expectations of human visitors. This is because technological advances continue to improve robots allowing them to find things people want and understand why people may want to visit a website including those things. As a result, the way your robotic visitors see your website will influence how many real people will eventually visit.

To help robots find your website, you can syndicate some of the pages using technologies like RSS feeds. These RSS feeds can distribute your pages to people and robots, helping them discover additions or edits to your website more quickly. These are especially helpful for frequently updated parts of your website, like news articles or blog posts. Similarly, you can distribute what's called an XML Sitemap as a feed to search engines directly. This file lists all of the pages, images, or videos contained on your website that you want search engines to find. While an XML Sitemap or another syndicated feed is no guarantee robots will find your website or send visitors to your website, it helps increase your chances, making it a worthwhile investment.

Once found, robots need to be able to access as much of your website as possible. You want robots to access every part of your website so that they can find everything and potentially use what they find to help attract visitors. However, some parts of your website should not be made public—like pages intended for members or subscribers. In these cases, you can place blocks within your code preventing robots from accessing. The blocks can take the form of suggestions contained on a text file—called a robots.txt file—about what parts of your website you want them to avoid. While the large majority of robots respect this file, some don't. So if a page contains sensitive information that should never be exposed publicly, stricter blocks can be included, like password protecting a sensitive file so that only those authorized can access it

You also typically want to allow these robots to include parts of your website in what they show about your website to potential visitors. For robots from search engines, you want to let them include text or images from your website in search results. Or, for robots from social networks, some of the text or images might be shown when a page from your website is shared within their network. The text or images these automated programs decide to show on the search result or in the social share acts as a teaser, giving people a reason to visit your website. The more the robot is allowed to include in these teasers, the more compelling your website will likely appear to potential visitors.

Finally, and arguably the more important piece, is ensuring robots understand the text you've included on your website. Many robots are tasked with scouting out your website in advance of human visitors. During a robot's reconnaissance mission to your website, you want it to clearly understand as much of your website as possible. The better the robots understand what your website offers and understands why a person would want to visit, the more likely it is they can help get humans who are interested in what your website offers to actually visit.

In the case of search engines, the better they understand exactly what products, services, or other information is included on your website, the better your chances your website will rank when people search for products, services, or information similar to what you offer. Much of the work of getting robots to understand your website depends on the nature of your text, images, and organizational structure. This is largely the same as the work done to help make your website understandable to humans. For instance, if your website's text uses the same words and phrases potential human visitors use, this makes it easier for robots to know where they should rank your website in their search results.

Technically, however, you can increase the chances of understandability by utilizing proper code structure within your website. Since this isn't a book about how to code, only two brief examples will be provided. First, by including the main header on a page or section in an <h1> tag, you indicate it's the first level or primary header. Secondary subheaders should be contained in a second-level header, or <h2> tag. Third-level headers would be in an <h3> tag (and so on). Using headers in this way helps robots understand the priority of text on your website as well as how the page is organized, which can have some influence on where those robots decide to rank your page within search results.[24]

As another example, robots are unable to see images in the same way as humans. To help robots know what is contained in an image, you can utilize alternative text providing a brief description of what the image contains.[25] This alternative text is provided in the alt attribute of the image tag. For instance, an image of your organization's logo might use an alt attribute with the text of your organization's name. This helps robots more clearly understand what your website's images are showing even if they can't "see" the image like human visitors do. As robots become clearer on what the images are, they can better determine where to show those images to human visitors. A robot from a search engine that finds your organization's logo could use the alt attribute containing the organization name as one way of determining if the image should appear within search results when a potential visitor searches for your organization.

Behavioral Considerations

. . .

Supporting Touch

Smartphones, tablets, and even some laptops are driven predominately by touch. Touch-based input gives people a new opportunity to physically interact with a website, and as a result, allows people to treat a website more like a physical object instead of a virtual object.[26] This physical interaction can make using a touch-based device more appealing and fun as it removes the barrier between the person and the screen created by a mouse and keyboard on a desktop or laptop computer.[27] This shift in interaction style alters the way people use websites and alters the way websites need to be built to support touch-based devices.

Tapping a link or button with a finger is, in many ways, similar to clicking that same link or button with a mouse. However, a finger can be used for other behaviors, like zooming or scrolling around a web page. The challenge is a finger is less precise than a mouse, and it's often less clear what people intended to do when tapping the screen—they may have intended to click but might have zoomed or scrolled instead. While the lack of precision makes it more challenging for people to use a website on a touch-based device, the finger-based input is easier to learn than using a mouse because it more closely mirrors natural behaviors.[28]

Where possible, you want to increase precision from touch input. One option is to alter the design of the things people need to tap—like links or buttons—to make sure they are large enough to touch. Also, the design of these tap targets needs to provide sufficient space around it so that it's clear exactly which item a visitor interacted with. The recommendation from Google and others is anything people need to tap should be at least 48 pixels tall and wide—on correctly configured designs this is about seven millimeters or just over a quarter of an inch.[29] Any smaller, and people will struggle to get their fingers on links, buttons, and other tap targets.

That is a minimum suggestion, but there is no harm in adding more space to the items you want people to tap. Studies have shown the time it takes a person to move toward a tap target, like a link or button, is directly connected with the distance to the object and the size of that object.[30] What this means is that any feature people need to interact with—like navigation links or call to action buttons— should be placed within easy reach and be large enough so that people can easily move their fingers toward and tap the object. A link that is too small will require a visitor to take more time to move their fingers toward the item and then tap it correctly.

When deciding how to place and size these components, you also want to consider how people's fingers are located when holding the smartphone. Many people use their smartphone with a single hand, navigating through a website with their thumbs. The area the thumb can reach—known as the thumb zone—is typically located near the bottom of the screen extending upward into the area people can reach with their thumbs without having to move their hands.[31] Anything placed outside this area will require people to adjust their grips or stretch their fingers in unnatural ways.

Complicating matters, about two-thirds of people use their right hands, while a third rely primarily on their left hands to browse.[32] This means the thumb zone will differ slightly for right- and left-handed visitors. There is an area at the bottom of the screen where these two thumb zones overlap, creating an ideal place for key links.

The thumb zone helps address one of the bigger challenges caused by visitors using different devices. Websites designed for larger screens typically have the navigation near the top of the screen. If the navigation isn't moved when that design is scaled down to a smaller screen, the navigation will be outside the thumb zone, making it harder for people to use.[33] So, if your smartphone visitors aren't using the navigation, consider moving the navigation into the thumb zone.

• • •
Mental Models

When people visit your website, they have expectations of how it will behave. These expectations are based on beliefs about how they think website works and expectations about how your website can help them get whatever it is they are seeking. These beliefs are usually different than the reality of how your website works as people rarely, if ever, take the time to intentionally study how your website performs to consciously update their beliefs and expectations.[34]

To an extent, these beliefs and expectations—called mental models—of how people predict a website will behave are based on other websites, which explains why consistency and following standards matters (as discussed in chapter 3). But this mental model is also shaped by how people think about things offline, in the nonvirtual world. This is why terms like "cart" can be used for online shopping and why various icons can help communicate concepts to visitors. Along with other websites and physical objects, the mental model can also be based on intuitive perceptions—also known as wild guesses—of how people think things should work.[35]

The challenge is each person visiting has a unique mental model of how your website should work. If every visitor shared the same mental model, designing and updating your website would be easy—once you understand what people believe and what their previous experiences are, then you can build to meet exactly those expectations. But each person has a wide array of past experiences, beliefs, and guesses about how websites behave. This means you aren't building your website for one mental model, but rather building to satisfy as many unique mental models as possible.

A further complication is that you, and the other people at your organization who manage the website, have a different mental model of how your website works than the people visiting. Unlike your visitors, your mental model more closely mirrors the actual behavior of your website. This is because you have used the website far more frequently and have deeper familiarity with the organization behind the website, which will greatly alter your past experiences and beliefs about how your website ought to behave. As well, with all of the information you know about your organization, your intuitive perceptions will probably be more accurate.

Your visitors, though, lack this deeper experience with your website and your organization. Because of that, you cannot rely on your mental model to understand a visitor's mental model. This is why visitors will rarely use the same words as you to describe the products, services, and other information your organization provides. This is also why visitors have different expectations for how your website should be organized.

To ensure your website fits the mental model of your visitors, you need to continually measure and monitor visitor activity to detect what people believe, what they expect to find, and what guesses they are making. You might notice people aren't clicking on a link that is critical for converting. This lack of a click might be because your visitors don't understand what the words in that link mean or they believe something different about what those words mean. By adjusting the wording, people may begin clicking and converting. An increase in clicks indicates the new words have better alignment with visitors' mental models.

As you consider what changes to make, you also want to monitor the sources people use to find your website. How people find your website, and the first things they see about your website, will shape their beliefs about what your website will contain. Because of this, there is a risk people will develop outdated or incorrect beliefs or expectations about your website.

Say, for example, some visitors might find a link to your website stating your organization carries a certain product. People who see this link will expect to find something about that product when they visit, whether you have that product in stock or not. By monitoring these sources, you can weed out old or incorrect links. Or you can find new ways to present the messages to visitors to let them know about the outdated or incorrect information. For instance, instead of telling visitors a product of interest is out of stock, why not also offer up a similar product that might meet their expectations?

Link by link, image by image, word by word, source by source, you want to tweak and adjust your website to match your visitors' mental models. While each visitor's mental model will be unique, you will start to find commonalities among the majority of your visitors. The more you can structure your website to meet these common mental models, the more people will see how engaging or converting with your website aligns with their beliefs and expectations.

What to Measure

• • •

Relevancy

The tasks people expect to complete during a visit to your website might include answering a question, browsing product, finding directions, or just killing time. Your organization has other types of tasks in mind, though, like selling products, generating leads, or getting people to attend an event. Chances are your organization's expectations won't completely match the expectations of visitors. Catering solely to visitor expectations while neglecting your organization's expectations might be satisfying for your visitors, but not very satisfying for your organization.

The goal then should be finding the intersection point between what visitors expect and what your organization expects. The more you can promote the tasks at this intersection point between the stuff people want to do on your website and the stuff your organization wants people to do, the more relevant your website will be to visitors. The more relevant a website is to visitors, the more people will see how engaging and converting will meet their expectations.

The first way to think of tasks—and measure tasks—is as existing across the entire website. With this view, the entire point of a visit to your website is people converting or engaging by completing a particular task. How many people placed an order last month? Or, how many people watched a video this week? These sitewide metrics also provide a high-level view of relevancy. A higher overall conversion or engagement rate suggests the things promoted in calls to action are closer to the intersection point. Visitors are getting whatever it is they wanted, so they engage and convert, which is also beneficial to your organization.

But along with looking a sitewide metrics like this, you also want to know if each individual page is relevant. A few pages might be extremely relevant, and these high-performing pages skew the overall engagement or conversion numbers higher. But other pages may be irrelevant to people visiting, causing people to leave before they ever see one of your website's high-performing pages. This means by improving those low-performing, irrelevant pages, you have an opportunity to make all aspects of your website relevant, which will make all of your pages perform better—resulting in your entire website performing better.

To understand the relevancy of each page, you want to measure engagements and conversions on a specific page. These are simple but meaningful tasks people complete while on that page of your website. On some pages, these tasks might be very similar to the broader, sitewide tasks—such as how many people added a product to a cart or how many people joined a newsletter. Other per-page tasks may be more page-centric, like how many people clicked a particular link to another page, scrolled to read more of the page's text, or spent more than a minute viewing the page.

By knowing the number of people who completed these per-page tasks, you can determine how well that page is meeting expectations. If the actions taken on a page are relatively low, the page isn't relevant—no matter how much those tasks or that page matters to your organization, people don't see how it helps them. But if many visitors are completing these tasks, that indicates they are benefiting from the page. As you make changes, you want to learn from the relevant pages to improve the irrelevant pages.

A common reason a page isn't relevant is that the expectations of the people visiting don't align with the expectations of your organization. You may wish people viewing a particular page complete the per-page task of adding a product discussed on that page to the shopping cart. But people visiting are only interested in browsing through pictures of the product or reading reviews before going elsewhere to place an order. Visitors are completing per-page tasks, so the reviews or pictures are relevant. The tasks visitors find relevant don't intersect with the tasks your organization thinks should be relevant. Perhaps by adding a better description, a better price, or a list of reasons your organization is superior to competitors, you may be able to show why they should buy this product from your organization instead. If successful, more people visiting this page may find adding the product to the cart to be relevant and in line with their expectations.

If those changes still don't get people purchasing the product, you may want to make a more drastic change by switching what you ask people to do on this page. People may just not find purchasing the product to be relevant to their needs, no matter how appealing your offer. Instead, the task that might be more relevant is getting people to call a sales rep for help reviewing the product they are about to purchase. Or maybe the change isn't pushing toward a sale at all but simply asking people to join your e-mail newsletter or connect on social media—both of which could still be valuable to your organization as well as to visitors.

Despite all the changes, it may be that nobody completes any task you would want them to complete; there is no intersection point. They don't want to buy your product or call your sales rep or join your newsletter or follow you social media. Instead, they just want to look at pictures, read reviews, and then go elsewhere, leaving your website never to return. In these cases where the engagement and conversion rates remain stubbornly low, the problem may not be with your page or the page's tasks. Instead, the problem may be you've attracted the wrong type of visitors.

Of course, there is nothing wrong with the people visiting your website, but these people will never find your website relevant simply because they don't want what your website has to offer. So along with measuring per-page tasks, you want to divide this metric one step further by measuring per-page tasks per source. You may find while task completions on a page are low overall, the per-page task completions are actually quite high for people arriving on this page from a link in a news article that reviewed your organization. The per-page task completions might be incredibly low, however, for visitors arriving from a Google search result or a share on Facebook. Based on this, you can alter your marketing to attract visitors from sources who are more likely to engage or convert.

This situation is an opportunity to find new sources to bring in visitors whose needs more closely intersect with your organization's needs. Or if you find certain sources send visitors who are more relevant, this is an opportunity to find ways of getting more visitors from those sources. Either way, this will result in more people completing the per-page tasks. As each individual page of your website becomes more relevant to what visitors want, your sitewide conversion and engagement rates will improve as well.

• • •

Matching Navigation to Mental Models

People's mental models—based on a mix of beliefs, guesses, and past experiences—inform what they think they'll get when they visit your website and also inform how they think they'll be able to get it. You need to offer people some means of navigating through your website after they've arrived. For the highest amount of engagement, your website needs to think like visitors think. Because visitors are most likely thinking in terms of topics first, this means your website's navigation needs to show the various topics that align with visitors' mental models. Mapping visitor mental models to navigation links requires answering many questions, including what links to show, should icons be used, what words or phrases to use, and what order links should be placed in.

These questions are particularly hard for new websites where it isn't clear who will visit the website. In the absence of interviewing a representative group of potential visitors, the best answer is to start by taking an educated guess on how to present your navigation. The biggest and most costly mistake many organizations make at this stage is relying on internal concepts about how the website should be organized and which links should be in the navigation. It's worth repeating again: you don't think like your visitors think.

So if interviews with your visitors aren't feasible, instead of relying on internal concepts about how the website should be organized, you can start instead by organizing your navigation on the topics visitors will be most interested in finding on your website. You can identify what these topics are, determine how to word these topics, and determine how to order the links by relying on best practices and the standards within your industry. At first, this can result in your website's navigational structure being similar to competitor websites.

But your website is unique and the reasons people visit will differ from why people visit other websites (otherwise, why would they visit your website?). The second largest mistake many organizations make is to not make changes to their websites' navigation. The navigation is among the most critical ways your website communicates with the people who are visiting. After your initial navigation is created, you can measure which topics people are most interested in and, more specifically, measure which specific links people use. This information will help you learn about your visitors' mental models. Using what you've learned, you can adjust your navigation to include more links to topics that are of greater interest.

A sophisticated approach to measuring navigation link usage is to set up tracking in analytics tools telling you precisely how many times each navigation item was clicked or tapped. The precise numbers are helpful, but a simpler option is to use a heatmap tool, which provides a visual report on which links people click or tap—see chapter 2 for more information about heatmaps. Along with looking at overall click or tap activity, you can also segment analytics tools or heatmap tools to show this activity in your website's navigation for visitors from a particular source. Like with measuring relevancy, you may find people visiting from one source—like Google or Facebook—might be more likely to use your navigation links than other sources.

As you review the measurements on where people have engaged

with the navigation, you might find, as an example, people rarely click on the links included in the top navigation bar, but many people do click on the links in the sidebar navigation. This might mean you have excluded items from the top navigation bar people were interested in finding. With this type of result, you could add the more popular links from the sidebar navigation into the top navigation bar.

Alternatively, the difference in clicks between the top and sidebar navigation might indicate something about the style of navigation you have used. Maybe the top navigation is organized following your organization's internal structure, but the sidebar navigation uses tasks or topics as the link text. If this is the case, it provides a clear reason to change the style used in the top navigation bar.

As another example, the differences between what is and is not being clicked may have more to do with the types of words used within the links. Maybe links in the top navigation bar contain broad terms while the links to the exact same pages in the sidebar use more specific terminology. In this example, visitors might find the specific terms more in line with their expectations. Similarly, the links in the top navigation might be written for people with a higher level of education than the words used in the sidebar. One of the best things to test within your website's navigation are different phrases you could use to link to any given page—some words or phrases will naturally better match visitors' expectations and thought patterns.

Of course, the difference in clicks between the top and sidebar navigation in this example could have nothing to do with the words or navigation style used, but instead have everything to do with the design. Perhaps people have an easier time seeing the link in the sidebar instead of in the top navigation bar because the font size of links happens to be larger in the sidebar or maybe the colors make the sidebar more prominent. By adjusting the design of the links in the top navigation bar, you could make this part of your website easier to use, increasing the number of people who use the links in the top navigation bar.

Supporting Measures

• • •

Word and Term Usage

To help people meet their expectations during a visit to your website requires speaking the visitor's language. This helps you match people's mental models and also increases the chances your website's text will be more relevant and interesting to your visitors. To begin, though, you need to research the various words or phrases your visitors actually use.

One way to research these words is by conducting interviews with the people likely to visit your website. If this is something that can be afforded, and time allows, it's an effective way to understand visitors' mental models and the language they are likely to use. But interviews aren't always practical for new organizations that are unsure of who their visitors will be or for projects with tight budgets or schedules.

An alternative way of finding the words your visitors use is with a keyword research tool. These tools let you find the words or phrases—or keywords—that are related to your industry people are typing into search engines. The words and phrases people type into a search engine are usually the same words they'd use when discussing or thinking about the products, services, or information your organization provides. Along with returning all the various keywords people use, these tools also provide at least approximate metrics on how many searches are conducted for the keyword over a given time range. The more searches there are for a keyword, the more the keyword is likely representative of your visitors' mental models, and the more likely it is you should consider using it in your text.

Before making a final decision on which keywords to use in your website's text or navigation, you want to confirm what people think when they search for that phrase. One way to do this is to conduct a search on Google or another search engine to see what websites appear in the results for that keyword. Search engines work hard to return relevant results for each keyword searched. Because of this, the search results can give you an indication of what types of websites people will find relevant for the keyword. If the websites that appear in the results are similar in nature to your website, you probably have found the right words and phrases to use.

A downside to researching keywords used in search engines is the data available in the keyword research tools tends to only be for the more popular keywords, and even for those terms, the data provided about search volume can be distorted.[36] Search phrases receiving only a few searches per month are, at best, unreliably found. If you are serving a smaller niche audience, or if you are establishing a new type of product, finding relevant keywords might prove difficult.

Another approach is to review the social media channels of people who are at least somewhat representative of the kinds of people you would like to visit your website. To the extent these people discuss your products or services, or something like your products or services, you can review what kinds of words or phrases they use. Unlike researching words used in search engines where many tools exist to streamline this process, reviewing social media posts from potential visitors is typically a manual process. But even reviewing a few social media posts from would-be visitors can help shift the way your text is written to more closely mirror how visitors think. If potential visitors to your website are not active on social media—or not active about what your organization does—you can also review e-mails from potential website visitors—such as e-mails from old customers or business partners—to see what words were used.

After researching the words and phrases your visitors use, you want to compile the words you've found into a dictionary. This way, you know what words and phrases to use, making maintaining consistency of terminology far easier. This dictionary can also be updated as you measure visitor activity. For instance, if you find nobody clicks or taps on certain words in your navigation, this can be noted to inform and improve future changes to the navigation. A dictionary also lets you detect patterns about your visitors to learn more about their mental models. If most of the words are simplistic and indicate a lack of knowledge about your industry, it's reasonable to assume that no expert-oriented terms should ever be used—even if a keyword research tool suggests otherwise.

* * *

Readability

People won't get what they want from a visit to your website if they struggle to read your website's text. Many tools exist to review your website's readability. These tools also tell you the grade level a visitor would need to reach in order to understand your text. The assessments—many of which are free—are based on a variety of different formulas and calculations, the more popular of which are the Flesch–Kincaid and Gunning Fog formulas. Each readability assessment tool is slightly different, so instead of relying on a single tool, the better approach is to review your website with several. This way you get a more complete evaluation of how readable your text is.

The question though is what grade level or readability score is appropriate for your visitors? To make this determination, you want to compare a page's grade level and readability score to the page's engagement and conversion rates. Pages with certain readability scores will also often have higher engagement or conversion rates— depending on your visitors, this may be a higher or lower grade level. Similarly, you could also review readability in comparison to metrics discussed in other chapters—like an error's continuation rate, time spent on the page, scrolling amounts, or bounce rates. These comparisons help you understand what your visitors want from your text and how you should write for your visitors going forward.

Another approach to determining how to make your text readable is to review the readability of competitor websites or other websites serving a similar purpose to yours. This can be especially helpful when trying to decide how to write text for a new website. While you don't want to directly copy whatever a competitor website has done, reviewing other websites to find what they have in common can give you an idea of where to begin. For instance, if most competitor websites write to a higher education level then perhaps your website should too.

As a word of warning, a competitor assessment assumes the people running those other websites were intentional in how they chose to write their text. This may not always be the case. For all you know, those other websites might have copied another website, who in turn copied another and so on. Despite this, using an external approach to determining how to write your website's text can give you a place to start developing your own website. But once your website is launched, refine the text using your own engagement and conversion rates to ensure you find the best way to speak to your visitors.

Evaluate Your Website

Language Expectations

- Do page titles, headers, links, calls to action, navigation, and the main text use words and phrases people would actually use when discussing whatever it is your organization does?

- Is the text written at an appropriate grade level for visitors? How does the grade level compare to engagement or conversion rates?

- If jargon or expert-oriented terms are used, are the terms used appropriately and in a way visitors would understand?

- If icons are used, is it clear what concepts the icons represent? Are icons used to support what text communicates instead of as the only representation of a piece of information?

Navigation and Calls to Action

- What type of navigation structure does your website use? Is it organized around internal structures, types of audiences, tasks people can complete, or broad topics of interest to visitors?

- What navigation items are clicked or tapped? What does this activity suggest about your visitors' mental models?

- Are conversions pushed for in places where people aren't interested in converting? Or are conversions offered in relevant locations where visitors have expressed an interest in converting?

Technical Expectations

- How quickly do the key pages on your website load? When clicking on links, submitting forms, placing orders, or other activities, does your website quickly demonstrate the cause and effect of a visitor's actions?

- For slower processes, are progress bars or other indicators used to provide a perception of speed?

- On touch-based devices, are links spaced far enough apart to allow links or buttons to be easily tapped?

- Are key links, buttons, calls to action, or navigation elements located in the thumb zone where people can easily tap?

Measurement Guide

For guidance setting up the tools for these measurements, see:
http://www.matthewedgar.net/elements/real

Baseline Setup	Define tasks that people want to complete on each page of your website and define the tasks your organization wants people to complete on each page. How do these tasks compare? The closer these two sets of tasks align, the higher your conversion and engagement rates. Research keywords your visitors would use when discussing that page's subject, either from search engines, social media, or interviews. Compile a dictionary of frequently used words, words to avoid, and general patterns of word or phrase use.
Monthly or Quarterly	What source lead people to your website? What do people from each source want and need? Are these sources aligned with what your website actually offers? Review click volume for each navigation link on your website. You can either use exact measurements or approximate click activity via a heatmap. Review sources and navigation clicks for visitors using different devices. Do visitors on smartphones use different sources or click on different links than visitors using desktop computers or tablets?
Before Major Changes	Check each page's readability score and compare to that page's engagement or conversion rates. If engagement or conversion rates are currently lower than you'd prefer, try changing the page's text to a higher or lower grade level.

FINAL THOUGHTS

IT'S EASY TO FEEL overwhelmed by all the elements on your website you could change. It's easy to think how awful your website currently is and the volume of work required to make improvements. It's easy for this to turn into procrastination, resulting in no changes being made.

Start small, start simple. Your goal is not perfection. Your goal is not to fix everything right now—no matter how tempting it might be. Your goal is making your website slightly better today than it was yesterday. Take five to ten minutes every day over the next month to make just one part of your website a little bit better. A month or two later, your website will work slightly better for the people visiting, resulting in more people engaging and that will lead to more people converting.

To decide what changes to make, start by reviewing the questions discussed in this book. What errors get in the way of people successfully visiting your website? Where are people making mistakes or slipping up? Is your text inconsistent, causing people to leave confused or dissatisfied? Does your design contain something abnormal or uncommon that causes frustration? What pages of your website are overly complicated and tedious? Are people coming to your website expecting something different than what you actually offer? Are people free to choose how to move about your website to find whatever it is they want to find, or is your website forcing people to do something they aren't interested in?

Measure the results of whatever you've changed so you know what works. Learn from what doesn't work and keep what does. Then repeat the process. It's these slow, steady, and continual changes that help you strike the right balance to meet the needs of your organization and the needs your visitors. The more changes you make, and the closer you get to achieving that balance, the more your website will succeed.

ADDITIONAL RESOURCES

Y OU NEED TO continually update your website because change is constant. But because change is constant, inevitably some of the specific examples in this book will become outdated. Technology advancements will lead to new things to consider and new questions to ask. The links in the Measurement Guides (copied below for convenience) contain additional resources to help you make your website a success.

While the specifics might change, the overall themes covered won't. These themes apply to the earliest websites launched in the 1990s and will apply to futuristic websites too—even if those futuristic websites are delivered via virtual reality, a chip in a person's head, voice-based devices, or some as-of-yet unimagined technology. Regardless of the technology used to visit, the people who visit your organization's website will need specific things, and they will want to control how they get those things instead of having something pushed on them. People will always be more likely to engage and convert if the website is simple, efficient, error-free, behaves consistently, speaks their language, and thinks the way they think.

Links to More Resources

- Simple and Efficient to Use:
  http://www.matthewedgar.net/elements/simple

- Allow Control and Offer Guidance:
  http://www.matthewedgar.net/elements/control

- Maintain Consistency and Follow Standards
  http://www.matthewedgar.net/elements/standards

- Prevent and Handle Errors:
  http://www.matthewedgar.net/elements/errors

- Real World, Real People:
  http://www.matthewedgar.net/elements/real

NOTES

Chapter 1: Simple and Efficient

1. Ada Ivanoff, "Design Minimalism: What, Why & How," June 6, 2014, https://www.sitepoint.com/what-is-minimalism/.
2. Kevin Mark Rabida, "The Difference Between Minimalism and Simplicity," February 23, 2015, http://www.ucreative.com/articles/minimalism-simplicity-difference/.
3. Laura Franz, "Chunking Text with Hierarchy," February 13, 2016, http://typographicwebdesign.com/setting-text/chunking-text-with-hierarchy/.
4. Oli Gardner, "How to Optimize Contact Forms for Conversions," April 11, 2013, http://unbounce.com/conversion-rate-optimization/how-to-optimize-contact-forms/.
5. Danny Halarewich, "Single-Page or Multi-Page Checkout: Which Is Better?," *LemonStand*, March 11, 2015, http://blog.lemonstand.com/the-great-debate-single-or-multi-page-e-commerce-checkout/.
6. Clicktale, "How Website Visitors Scroll and See Your Content," *Clicktale*, December 4, 2007, https://www.clicktale.com/academy/blog/are-your-web-visitors-really-paying-attention/.
7. Clicktale, "Scrolling Research Report V2.0—Part 1: Visibility and Scroll Reach," *Clicktale*, October 5, 2007," https://www.clicktale.com/academy/blog/clicktale-scrolling-research-report-v20-part-1-visibility-and-scroll-reach/.
8. Google, "The Importance of Being Seen," November 2014, http://think.storage.googleapis.com/docs/the-importance-of-being-seen_study.pdf.
9. Unbounce, "Call to Action Design | Landing Page Conversion | Part 3," 2016, http://thelandingpagecourse.com/call-to-action-design-cta-buttons/.
10. Amy Schade, "The Fold Manifesto: Why the Page Fold Still Matters," *Nielsen Norman Group*, February 1, 2015, https://www.nngroup.com/articles/page-fold-manifesto/.
11. Ott Niggulis, "Which Color Converts the Best?," *ConversioXL*, January 16, 2013, http://conversionxl.com/which-color-converts-the-best/.

12. Tom Osborne, "Color Contrast for Better Readability," *Viget*, March 3, 2015, https://www.viget.com/articles/color-contrast; Dmitry Fadeyev,"9 Common Usability Mistakes in Web Design," *Smashing Magazine*, February 18, 2009, https://www.smashingmagazine.com/2009/02/9-common-usability-blunders/.

13. W3C, "Contrast (Minimum) Understanding Success Criterion 1.4.3," *Understanding WCAG 2.0*, November 10, 2016, https://www.w3.org/TR/UNDERSTANDING-WCAG20/visual-audio-contrast-contrast.html.

14. To calculate color contrasts, see: WebAIM, "WebAIM: Color Contrast Checker," http://webaim.org/resources/contrastchecker.

15. Tara Hornor,"10 Troublesome Colors to Avoid in Your Advertising," May 8, 2013, https://www.sitepoint.com/10-troublesome-colors-to-avoid-in-your-advertising/.

16. Colors On The Web, "Color Contrast," http://www.colorsontheweb.com/Color-Theory/Color-Contrast.

17. National Eye Institute, "Facts about Color Blindness," February 2015, https://nei.nih.gov/health/color_blindness/facts_about.

18. Janet M. Six, "Designing for Senior Citizens | Organizing Your Work Schedule," May 17, 2010, http://www.uxmatters.com/mt/archives/2010/05/designing-for-senior-citizens-organizing-your-work-schedule.php.

19. Tom Brinck et al., *Usability for the Web: Designing Web Sites That Work* (San Francisco, CA: Morgan Kaufmann, 2001), 59.

20. Jeff Johnson, *Designing with the Mind in Mind: A Simple Guide to Understanding User Interface Design Rules* (Burlington, MA: Morgan Kaufmann Publishers/Elsevier, 2010), 62.

21. Jared Smith, "WCAG 2.0 and Link Colors," *WebAIM*, July 24, 2009, http://webaim.org/blog/wcag-2-0-and-link-colors/.

22. Olivier Thereaux and Susan Lesch, "Care with Font Size," April 9, 2010, https://www.w3.org/QA/Tips/font-size.

23. Patrick Sexton, "Use Legible Font Sizes," March 5, 2016, http://varvy.com/mobile/legible-font-size.html; D. Bnonn Tennant, "16 Pixels: For Body Copy. Anything Less Is A Costly Mistake," *Smashing Magazine*, October 7, 2011, https://smashingmagazine.com/2011/10/16-pixels-body-copy-anything-less-costly-mistake/.

24. OSHA, "Workstation Components Monitors," February 4, 2015, https://osha.gov/SLTC/etools/computerworkstations/components_monitors.html.

25. Mark Rosenfield, "Computer Vision Syndrome: A Review of Ocular Causes and Potential Treatments," *Ophthalmic and Physiological Optics* 31, no. 5 (2011): 504.

26. iA Inc., "The 100% Easy-2-Read Standard," November 17, 2006, https://ia.net/know-how/100e2r.

27. Google Developers, "Use Legible Font Sizes," April 8, 2015, http://developers.google.com/speed/docs/insights/UseLegibleFontSizes.

28. Ilene Strizver, "Serif vs. Sans for Text in Print," April 2013, https://www.fonts.com/content/learning/fontology/level-1/type-anatomy/serif-vs-sans-for-text-in-print.

29. Carrie Cousins, "Serif vs. Sans Serif Fonts: Is One Really Better Than the Other?," October 28, 2013, https://designshack.net/articles/typography/serif-vs-sans-serif-fonts-is-one-really-better-than-the-other/.

30. Mind Tools Editorial Team, "Cognitive Load Theory: Making Learning More Effective," April 9, 2016, https://www.mindtools.com/pages/article/cognitive-load-theory.htm.

31. John Sweller et al., *Cognitive Load Theory* (New York: Springer, 2011), 68.

32. Roland Brünken, *Cognitive Load Theory: Theory and Applications* (Cambridge: Cambridge University Press, 2010), 134.

33. Johnson, *Designing with the Mind in Mind*, 111.

34. Ibid., 112.

35. Margherita Antona et al., "Universal Access in Human-Computer Interaction: Methods, Techniques, and Best Practices: 10th International Conference," Held as Part of HCI International 2016, Toronto, ON, Canada, July 17–22, 2016, Proceedings. Part I. n.p., 2016, 479.

36. For example, see ISO 9241-11, "Ergonomic Requirements for Office Work with Visual Display Terminals: (VDTs)—Part 11: Guidance on Usability." March 19, 1998, Last Modified January 23, 2013, http://www.iso.org/iso/catalogue_detail.htm?csnumber=16883.

37. Chris Miksen, "What Is the Difference Between Efficiency and Effectiveness in Business?," *Chron* (Chron.com), December 2011, http://smallbusiness.chron.com/difference-between-efficiency-effectiveness-business-26009.html.

38. Avinash Kaushik, "Standard Metrics Revisited: #3: Bounce Rate," August 6, 2007, http://kaushik.net/avinash/standard-metrics-revisited-3-bounce-rate/.

Chapter 2: Allow Control and Offer Guidance

1. Arnold M. Lund, "Expert Ratings of Usability Maxims," *Ergonomics in Design: The Quarterly of Human Factors Applications* 5, no. 3 (1997): 15–20, maxim 14.

2. New York Department of Labor, "User Control Freedom—User Experience Guide," accessed November 18, 2016, http://www.labor.ny.gov/ux/principles-user-control-freedom.html.

3. Ross Campbell et al., "User Control and Freedom," 2010, http://mattsoave.com/old/cogs187a/iu_site_eval/3control.html.

4. Arnold M. Lund, "Expert Ratings of Usability Maxims," maxim 25. In his heuristics, Jakob Nielsen refers to this concept as an "emergency exit."

5. Mike Fisher, "6 Ways to Improve Confirmation Pages," March 19, 2009, http://completeusability.com/6-ways-to-improve-confirmation-pages/.

6. Neil Patel, "9 Landing Page Elements That Need to Die," *The Daily Egg*, July 15, 2014, https://blog.crazyegg.com/2014/07/15/remove-landing-page-elements/.

7. Diana Urban, "Should You Remove Navigation from Your Landing Pages? Data Reveals the Answer," January 9, 2014, http://blog.hubspot.com/marketing/landing-page-navigation-ht.

8. Amy Schade, "Don't Force Users to Register Before They Can Buy," *Nielsen Norman Group*, July 15, 2015, https://www.nngroup.com/articles/optional-registration/.

9. Anthony, "Why Hover Menus Do Users More Harm Than Good," *UX Movement*, March 1, 2011, http://uxmovement.com/navigation/why-hover-menus-do-users-more-harm-than-good/.

10. Matt Cronin, "Designing Drop-down Menus: Examples and Best Practices," *Smashing Magazine*, March 24, 2009, https://smashingmagazine.com/2009/03/designing-drop-down-menus-examples-and-best-practices/.

11. Jakob Nielsen, "Drop-down Menus: Use Sparingly," *Nielsen Norman Group*, November 12, 2000, https://www.nngroup.com/articles/drop-down-menus-use-sparingly/.

12. Katie Sherwin, "The Magnifying-Glass Icon in Search Design: Pros and Cons," *Nielsen Norman Group*, February 23, 2014, https://www.nngroup.com/articles/magnifying-glass-icon/.

13. Kara Pernice and Raluca Budiu, "Hamburger Menus and Hidden Navigation Hurt UX Metrics," *Nielsen Norman Group*, June 26, 2016, https://www.nngroup.com/articles/hamburger-menus/.

14. James Foster, "Mobile Menu AB Tested: Hamburger Not the Best Choice?," September 27, 2016, http://exisweb.net/mobile-menu-abtest.

15. Arnold M. Lund, "Expert Ratings of Usability Maxims," maxim 33.

16. For a way of handling this technically, see: https://davidwalsh.name/change-text-size-onclick-with-javascript.

17. Each browser handles this in different ways. See: Firefox (https://support.mozilla.org/en-US/kb/font-size-and-zoom-increase-size-of-web-pages); Chrome (https://support.google.com/chrome/answer/96810?hl=en); Internet Explorer (https://support.microsoft.com/en-us/help/17456/windows-internet-explorer-ease-of-access-options).

18. To prevent this chapter from becoming a lengthy, technical discussion of relative units, only a very brief summary was included. For more details on these units and how they compare, see: https://zellwk.com/blog/rem-vs-em/ or https://www.futurehosting.com/blog/web-design-basics-rem-vs-em-vs-px-sizing-elements-in-css/.

19. Google Developers, "Size Content to Viewport," April 8, 2015, accessed July 19, 2016, http://developers.google.com/speed/docs/insights/SizeContentToViewport.

20. You can disable zooming for visitors on mobile devices by setting the "user-scalable" attribute to "0" or "no," but there are few use cases where this makes sense.

21. US General Services Administration, Section 508 Reference Guide, 1194.22 (a).

22. For a discussion on these techniques, see: https://css-tricks.com/responsive-images-youre-just-changing-resolutions-use-srcset/.

23. Jakob Nielsen, "10 Best Application UIs," *Nielsen Norman Group*, August 12, 2008, https://www.nngroup.com/articles/10-best-application-uis/.

24. Phan, "Helping Users Easily Access Content on Mobile," Google Webmaster Central Blog, August 23, 2016, https://webmasters.googleblog.com/2016/08/helping-users-easily-access-content-on.html.

25. Danny Richman, "How Online Visitors Feel about Pop-Ups. How Not to Annoy Your Customers," July 20, 2016, https://www.seotraininglondon.org/should-use-popups-website/.

26. George Mathew, "Opt-in Pop-Ups: Are They Any Good?," *The Daily Egg*, August 18, 2014, https://blog.crazyegg.com/2014/08/18/opt-pop-ups/.
27. Ott Niggulis "In Defense of the Email Popup," *ConversionXL*, June 26, 2014, http://conversionxl.com/popup-defense/.
28. Brian Massey, "7 Best Practices for Your Exit-Intent Popovers and Popups," *Conversion Sciences*, January 20, 2015, http://conversionsciences.com/blog/7-best-practices-using-exit-intent-popovers/.
29. For more about the psychology of offering too much choice, see Barry Schwartz's 2004 book *Paradox of Choice*.
30. Stefanie Grieser, "Is Too Much Choice Killing Your Conversion Rates?," April 10, 2014, http://unbounce.com/conversion-rate-optimization/psychology-of-choice-conversion-rates/.
31. Barry Schwartz, "Is the Famous 'Paradox of Choice' a Myth?" PBS (PBS NewsHour), February 20, 2014, http://pbs.org/newshour/making-sense/is-the-famous-paradox-of-choic/.
32. Jason Fried, "Ask 37signals: Is It Really the Number of Features That Matter?," October 10, 2007, https://signalvnoise.com/posts/643-ask-37signals-is-it-really-the-number-of-features-that-matter.
33. Chris Anderson, *The Long Tail: Why the Future Is Selling Less of More*. Revised and Updated Edition (New York: Hyperion Books, 2008), 174.
34. The theory was initially proposed by Pirolli and Card in the article "Information Foraging" published in the Psychological Review in 1999. For more information, you can read a detailed summary of information foraging at https://www.interaction-design.org/literature/book/the-glossary-of-human-computer-interaction/information-foraging-theory.
35. Jakob Nielsen, "Information Foraging: Why Google Makes People Leave Your Site Faster," *Nielsen Norman Group*, June 30, 2003, http://nngroup.com/articles/information-scent/.
36. James Foster, "Mobile Menu AB Tested," http://exisweb.net/mobile-menu-abtest.
37. Avinash Kaushik, "Excellent Analytics Tip #13: Measure Macro AND Micro Conversions," March 26, 2008, http://www.kaushik.net/avinash/excellent-analytics-tip-13-measure-macro-and-micro-conversions/.
38. Himanshu Sharma, "MSc Digital Marketing Online," November 6, 2012, http://www.webanalyticsworld.net/2012/11/how-to-separate-user-engagement-from-distraction.html.

39. What matters more is the pixel density of the visitors screen, but the general idea applies that more width and height available, the more you can show to a visitor. See: http://www.tested.com/tech/371-why-pixel-density-matters-more-than-just-screen-size-or-resolution/.

40. Pawel Piejko, "Most Popular Smartphone Screen Resolutions in 2015," July 2, 2015, https://deviceatlas.com/blog/most-popular-smartphone-screen-resolutions-2015.

41. Kurt Maine, "List of Tablet and Smartphone Resolutions and Screen Sizes," November 10, 2011, http://binvisions.com/articles/tablet-smartphone-resolutions-screen-size-list/.

42. RapidTables, "Screen Resolution Statistics," February 2014, http://www.rapidtables.com/web/dev/screen-resolution-statistics.htm.

43. Though some studies show most people hold a phone vertically. See: http://www.uxmatters.com/mt/archives/2013/02/how-do-users-really-hold-mobile-devices.php.

44. Jason Pamental, "A More Modern Scale for Web Typography," January 15, 2015, http://typecast.com/blog/a-more-modern-scale-for-web-typography.

45. Mon Chu Chen et al., "What Can a Mouse Cursor Tell Us More? Correlation of Eye/Mouse Movements on Web Browsing." In *CHI '01 Extended Abstracts on Human factors in computing systems*," pp. 281–82, ACM, 2001; Jeff Huang et al., "User See, User Point: Gaze and Cursor Alignment in Web Search," in *Proceedings of the SIGCHI Conference on Human Factors in Computing Systems*, pp. 1341–50, ACM, 2012.

Chapter 3: Maintain Consistency and Follow Standards

1. Gerry Gaffney, "Why Consistency Is Critical," February 25, 2005, https://www.sitepoint.com/why-consistency-is-critical/.

2. Kathryn Whitenton, "Website Logo Placement for Maximum Brand Recall," *Nielsen Norman Group*, February 21, 2016, https://nngroup.com/articles/logo-placement-brand-recall/.

3. Jakob, Nielsen, "Horizontal Attention Leans Left," *Nielsen Norman Group*, April 6, 2010, https://www.nngroup.com/articles/horizontal-attention-leans-left/.

4. Plenty of exceptions to the left-positioned logo exist. See http://tomkenny.design/articles/the-use-of-logos-in-web-design/.

5. Code My Views, "Website Navigation Trends: 10 Tips & Examples," *Code My Views*, September 14, 2015, Last Modified July 31, 2016, https://codemyviews.com/blog/website-navigation-trends-10-tips-examples.

6. Jerry Cao, "The Hidden Power of Inconsistency in Design," *The Next Web*, June 30, 2015, http://thenextweb.com/dd/2015/06/30/the-hidden-power-of-inconsistency-in-design/.

7. Doug Kirk, "Embrace the Controversial: Why You Should Publish Pricing on Your Website," July 25, 2013, http://blog.hubspot.com/marketing/why-publish-pricing-on-website-var.

8. There is a lot to consider when adding a price to your website. For more, see https://stickybranding.com/should-you-publish-pricing-on-your-website/ and http://passionforbusiness.com/blog/should-you-put-your-prices-on-your-website/.

9. Google, "The New Multi-screen World: Understanding Cross-platform Consumer Behavior," August 2012, 20.

10. Ibid., 21.

11. Ibid., 22.

12. Brad Frost, "Content Parity," March 22, 2012, http://bradfrost.com/blog/mobile/content-parity/.

13. Google's recommended approach to designing for mobile is using responsive design techniques. http://developers.google.com/webmasters/mobile-sites/mobile-seo/.

14. Daniel Wesley, "How the World Spends Its Time Online," June 16, 2010, https://creditloan.com/blog/how-the-world-spends-its-time-online/.

15. Saul McCleod, "Cognitive Dissonance," 2014, http://www.simplypsychology.org/cognitive-dissonance.html.

16. Eric Wargo, "How Many Seconds to a First Impression?," July 1, 2006, https://psychologicalscience.org/observer/how-many-seconds-to-a-first-impression.

17. Danny Gallagher, "Goldfish Have a Better Attention Span Than You, Smartphone User," May 18, 2015, http://www.cnet.com/news/goldfish-the-actual-fish-not-the-crackers-may-have-a-better-attention-span-than-humans/.

18. Javier Bargas-Avila, "Users Love Simple and Familiar Designs—Why Websites Need to Make a Great First Impression," *Google Research Blog*, August 29, 2012, http://research.googleblog.com/2012/08/users-love-simple-and-familiar-designs.html.

19. Creative Bloq, *Why Visual Consistency Can Make or Break Your Web Design*, May 22, 2015, http://www.creativebloq.com/web-design/why-visual-consistency-can-make-or-break-your-web-design-51514942.
20. Google, "The New Multi-screen World," 25.
21. Ibid., 28.
22. Art Markman, "How Distraction Can Disrupt You," *Psychology Today*, March 18, 2014, http://psychologytoday.com/blog/ulterior-motives/201403/how-distraction-can-disrupt-you.
23. Jeff Johnson, *Designing with the Mind in Mind: A Simple Guide to Understanding User Interface Design Rules* (Burlington, MA: Morgan Kaufmann Publishers/Elsevier, 2010), 122.
24. Peep Laja, "PXL: A Better Way to Prioritize Your A/B Tests," *ConversionXL*, September 27, 2016, http://conversionxl.com/better-way-prioritize-ab-tests/.
25. Tracey Wallace, "Why Your Ecommerce Conversion Rate Determines Your Site's Success," *Big Commerce*, April 1, 2016, https://bigcommerce.com/blog/conversion-rate-optimization/.
26. To calculate this, see: https://vwo.com/ab-split-test-duration/ or https://www.optimizely.com/resources/sample-size-calculator/.
27. Rich Page, "A/B, Split and Multivariate Test Duration Calculator," *Visual Website Optimizer*, July 13, 2016, http://rich-page.com/website-optimization/how-to-test-and-improve-your-website-if-your-traffic-is-too-low-for-ab-testing/.
28. Josh Clark, *Tapworthy: Designing Great IPhone Apps* (Sudbury, MA: O'Reilly Media, Inc., 2010), 32; Jakob Nielsen, "Mobile Content: If in Doubt, Leave It Out," *Nielsen Norman Group*, October 10, 2011, http://nngroup.com/articles/condense-mobile-content/.
29. Larry Marine, "Responsive Design Vs. Task-Oriented UX Design," *Search Engine Watch*, February 27, 2014, https://searchenginewatch.com/sew/how-to/2331203/responsive-design-vs-task-oriented-ux-design.
30. Karen McGrane, "Your Content, Now Mobile," *A List Apart*, November 5, 2012, http://alistapart.com/article/your-content-now-mobile.

Chapter 4: Prevent and Handle Errors

1. Adapted from Jeff Sauro, "Rating the Severity of Usability Problems: MeasuringU," *MeasuringU*, July 30, 2013,

http://measuringu.com/blog/rating-severity.php and Jakob Nielsen, "Severity Ratings for Usability Problems: Article by Jakob Nielsen," *Nielsen Norman Group*, January 1, 1995, http://nngroup.com/articles/how-to-rate-the-severity-of-usability-problems/.

2. Don A. Norman, *The Design of Everyday Things*. Revised and Expanded edition (New York: Basic Books, 2013), 170–71.

3. Jakob Nielsen, "Error Message Guidelines," *Nielsen Norman Group*, June 24, 2001, https://www.nngroup.com/articles/error-message-guidelines/.

4. Thomas Fuchs, "How to Write a Great Error Message," *Medium*, August 30, 2015, https://medium.com/@thomasfuchs/how-to-write-an-error-message-883718173322.

5. Nick Babich, "Mobile UX Design: User Errors," June 3, 2016, http://babich.biz/mobile-ux-design-user-errors/.

6. Anthony, "How to Make Your Form Error Messages More Reassuring," *UX Movement*, August 23, 2013, http://uxmovement.com/forms/how-to-make-your-form-error-messages-more-reassuring/; Luke Wroblewski, "Inline Validation in Web Forms," *A List Apart*, September 1, 2009, http://alistapart.com/article/inline-validation-in-web-forms.

7. Arnold M. Lund, "Expert Ratings of Usability Maxims," *Ergonomics in Design: The Quarterly of Human Factors Applications* 5, no. 3 (1997): 15–20, maxim 16.

8. Marina Lebed, "Best Practice for Proactive Chat: To Engage or Not Engage," *Provide Support Blog*, October 5, 2013, http://www.providesupport.com/blog/best-practice-for-proactive-chat-to-engage-or-not-engage/; Anna Cheung, "8 Proactive Chat Best Practices with Ready-to-Use Scripts- Comm100 Blog," *Comm 100*, February 21, 2016, http://comm100.com/blog/proactive-chat-best-practices.html.

9. Janko Jovanovic, "Web Form Validation: Best Practices and Tutorials," *Smashing Magazine*, July 7, 2009, https://www.smashingmagazine.com/2009/07/web-form-validation-best-practices-and-tutorials/.

10. Forms that work, "Appearance: Labels, Buttons, Required Fields, and Making Forms Look Good," December 22, 2014, http://formsthatwork.com/Appearance; Derek Featherstone, "Required Form Fields," *Simply Accessible*, October 5, 2005, http://simplyaccessible.com/article/required-form-fields/.

11. Sören Preibusch et al., "The Privacy Economics of Voluntary Over-Disclosure in Web Forms," in *The Economics of Information Security and Privacy*, ed. Rainer Böhme (Berlin, Heidelberg: Springer, 2013), 183–209.

12. The frontend and backend code can work together in various ways other than these methods. As well, frontend code can be used to check for errors after a visitor submits the form. The example provided tends to be more common, and allows for an illustration of the bigger point.

13. Vinay Shet, "Are You a Robot? Introducing 'No CAPTCHA reCAPTCHA,'" *Google Security Blog*, December 3, 2014, https://security.googleblog.com/2014/12/are-you-robot-introducing-no-captcha.html.

14. Elie Bursztein et al., "How Good Are Humans at Solving CAPTCHAs? A Large Scale Evaluation," in *IEEE Symposium on Security and Privacy*, pp. 399–413, 2010.

15. Tim Allen, "Having a CAPTCHA Is Killing Your Conversion Rate," *Moz*, August 5, 2013, https://moz.com/blog/having-a-captcha-is-killing-your-conversion-rate.

16. Elisa Silverman, "Do the New Captchas Affect Conversion Rate?," *Pagewiz*, February 10, 2016, http://www.pagewiz.com/blog/conversion-rate-optimization/new-captchas-conversion-rate.

17. Patrick Phillips, "The Entire World Hates CAPTCHA. So Why's It Still Here?," *Patrick's Place*, July 4, 2011, http://patrickkphillips.com/blogging/entire-world-hates-captcha-whys-still/.

18. Brad Miller, "404 Page Best Practices," *Search Engine Watch*, September 11, 2013, https://searchenginewatch.com/sew/how-to/2293339/404-page-best-practices.

19. MarketingExperiments, "Friction: Are Your Webpages Rubbing Customers the Wrong Way?," *MarketingExperiments*, July 2, 2007, http://marketingexperiments.com/blog/research-topics/site-design/friction-are-your-webpages-rubbing-customers-the-wrong-way.html.

20. Jeremy Smith, "How to Win the War Against Conversion Friction," *Kissmetrics blog*, June 10, 2014, https://blog.kissmetrics.com/war-against-conversion-friction/.

21. John Tackett, "Landing Page Optimization: What a 29% Drop in Conversion Can Teach You about Friction," *Marketing Experiments*, July 7, 2014, http://marketingexperiments.com/blog/research-topics/landing-page-optimization-research-topics/drop-in-conversion-teach-friction.html.

22. Usability First, "Usability Glossary - Graceful Flow," May 11, 2010, http://www.usabilityfirst.com/glossary/graceful-flow/.

23. Google, "Create Useful 404 Pages," http://support.google.com/webmasters/answer/93641; Joost De Valk, "404 Error Page for WordPress—A Practical Guide," *Yoast*, November 3, 2009, https://yoast.com/dev-blog/404-error-pages-wordpress/.

24. Jessica Enders, "Well-Designed Error Messages," *Formulate Information Design*, January 11, 2011, http://formulate.com.au/blog/well-designed-error-messages.

25. Don A. Norman, *The Design of Everyday Things*, 171; Page Laubheimer, "Preventing User Errors: Avoiding Unconscious Slips," *Nielsen Norman Group*, August 23, 2015, http://nngroup.com/articles/slips/; Rachael Rettner, "How Your Brain Works on Autopilot," *Live Science*, June 9, 2010, http://livescience.com/6557-brain-works-autopilot.html.

26. Don A. Norman, *The Design of Everyday Things*, 170.

Chapter 5: Real World, Real People

1. Arnold M. Lund, "Expert Ratings of Usability Maxims," *Ergonomics in Design: The Quarterly of Human Factors Applications* 5, no. 3 (1997): 15–20, maxim 1.

2. One exception might be educational websites. But the text that gets people to initially engage with the educational material would likely need to be written in a language the beginners in need of the education offered would understand.

3. Impact Information Plain Language Services, "Know Your Readers," *Plain Language at Work Newsletter*, 2013, http://www.impact-information.com/impactinfo/literacy.htm.

4. Charles Arthur, "Phone Handsets: The Icon That, Like Floppy Drives, Will Not Die," *The Guardian*, August 23, 2012, http://theguardian.com/technology/blog/2012/aug/23/icons-floppy-phone-handset.

5. For more about the problems with using icons, see UX Myths article: http://uxmyths.com/post/715009009/myth-icons-enhance-usability.

6. Nathan Barry, "How to Use Icons to Support Content in Web Design," *Smashing Magazine*, March 3, 2009, http://www.smashingmagazine.com/2009/03/how-to-use-icons-to-support-content-in-web-design/.

7. Gerry McGovern, "Search Articles," *New Thinking,* January 29, 2012, http://www.gerrymcgovern.com/new-thinking/why-audience-navigation-usually-doesn%E2%80%99t-work.

8. Everyl Yankee, "Object-Focused Vs. Task-Focused Design," *UX Mastery,* December 7, 2012, http://uxmastery.com/object-focused-vs-task-focused/.

9. Qualaroo, "Reducing Bounce and Exit Rates," 2016, https://qualaroo.com/beginners-guide-to-cro/reducing-bounce-and-exit-rates/.

10. Jeremy Smith, "Optimizing For Conversions: Increasing Relevance," *Jeremy Said,* April 7, 2016, http://jeremysaid.com/blog/conversion-triangle-series-increasing-relevance/.

11. Jeff Johnson, *Designing with the Mind in Mind: A Simple Guide to Understanding User Interface Design Rules* (Burlington, MA: Morgan Kaufmann Publishers/Elsevier, 2010), 195.

12. Google Developers, "Mobile Analysis in PageSpeed Insights," April 8, 2015, https://developers.google.com/speed/docs/insights/mobile.

13. Akamai Technologies, "2 Seconds Is Threshold for ECommerce Performance," September 14, 2009, https://www.akamai.com/us/en/about/news/press/2009-press/akamai-reveals-2-seconds-as-the-new-threshold-of-acceptability-for-ecommerce-web-page-response-times.jsp.

14. Sean Work, "How Loading Time Affects Your Bottom Line," *Kissmetrics,* http://blog.kissmetrics.com/loading-time/.

15. Maile Ohye, "Site Performance For Webmasters," *Google Webmasters,* YouTube video, 12:10. Posted May 2010, 9:14–9:46.

16. Steve Souders, "The Performance Golden Rule," *SteveSouders.com,* February 10, 2012, http://stevesouders.com/blog/2012/02/10/the-performance-golden-rule/.

17. HTTP Archive Trends (as of November 2016), http://httparchive.org/trends.php?s=All&minlabel=Sep+15+2011&maxlabel=Nov+1+2016.

18. Sourav Kundu, "How Does WordPress Caching Work?," *WPExplorer,* October 28, 2013, http://www.wpexplorer.com/wordpress-caching-work/.

19. Ilya Grigorik, "HTTP Caching," *Web Fundamentals* (Google Developers), December 21, 2016, https://developers.google.com/web/fundamentals/performance/optimizing-content-efficiency/http-caching.

20. Matt West, "An Introduction to Perceived Performance," *Treehouse Blog*, April 23, 2014, http://blog.teamtreehouse.com/perceived-performance.
21. Chris Harrison et al, "Faster progress bars: manipulating perceived duration with visual augmentations," 1548.
22. See Google's support document discussing Googlebot: https://support.google.com/webmasters/answer/182072.
23. For example, see Facebook for Developers' guide to their crawler: http://developers.facebook.com/docs/sharing/webmasters/crawler.
24. Aleh Barysevich, "How Important Are Tags in 2016 for SEO?" https://www.searchenginejournal.com/important-tags-2016-seo/156440/.
25. Google, "Image Publishing Guidelines," https://support.google.com/webmasters/answer/114016?hl=en.
26. Josh Clark, *Tapworthy*, 56.
27. Jakob Nielsen, "Mouse Vs. Fingers as Input Device," http://www.nngroup.com/articles/mouse-vs-fingers-input-device/.
28. Ibid.
29. Google Developers, "Size Tap Targets Appropriately," http://developers.google.com/speed/docs/insights/SizeTapTargets Appropriately.
30. Mehmet Goktürk, "Fitts's Law," https://www.interaction-design.org/literature/book/the-glossary-of-human-computer-interaction/fitts-s-law.
31. Josh Clark, *Tapworthy*, 56–58; Eckert, "Mobile Web Design Patterns—A Look at the Thumb Zone," http://www.parachutedesign.ca/blog/mobile-design-patterns-a-look-at-the-thumb-zone.
32. Steven Hoober, "How Do Users Really Hold Mobile Devices?" http://www.uxmatters.com/mt/archives/2013/02/how-do-users-really-hold-mobile-devices.php.
33. Luke Wroblewski, "Responsive Navigation: Optimizing for Touch Across Devices," http://www.lukew.com/ff/entry.asp?1649.
34. Jakob Nielsen, "Mental Models and User Experience Design," https://www.nngroup.com/articles/mental-models/.
35. Susan Weinschenk, "The Secret to Designing an Intuitive UX," https://uxmag.com/articles/the-secret-to-designing-an-intuitive-user-experience.
36. Russ Jones, "Google Keyword Planner's Dirty Secrets," https://moz.com/blog/google-keyword-planner-dirty-secrets.

GLOSSARY

Alt Attribute. An alt (or alternative) attribute can be added to the HTML image tag and text can be assigned to this attribute. The text is commonly referred to as alt text and this text explains what the image contains to visitors who are unable to see the image.

Analytics Tool. An analytics tool collects and analyzes some type of data about how people use your website. Analytics can also be defined more broadly to go beyond the website to look at core business metrics, like revenue or brand reach.

Backend or Server-Side Code. Code for a website can reside in two places, one of which is on the server (also see frontend). This code has direct access to the server, including applications and databases. It is called backend because it is further away from the part of your website a visitor sees. Also see Frontend Code.

Bail Out. Some type of method that offers people a way to undo or reverse a mistake or an unwanted action.

Bounce Rate. The percentage of visits where a visitor looked at one page and did nothing else before leaving your website.

Blur Event. A blur event occurs when a visitor moves away from a form field. Typically, this happens when a visitor is done entering data into that field. Also see Focus Event.

Cognitive Load. A measure of how much mental effort people have to exert in order to use and understand your website.

Conversion. Conversions happen when visitors complete some type of important action on your website. Often, this action is also economically desirable for your organization such as getting an inquiry from a potential client, a donation, or a new subscriber.

CSS. CSS (Cascading Style Sheets) is a part of the frontend code and establishes rules for how various parts of your website should look. These rules can apply to standard HTML tags or custom features on your website.

Distraction Rate. This metric is a ratio of engagements or conversions and the time people spend on your website. If time increases but engagements or conversions decrease, chances are visitors are distracted (or lost), which makes them unable to convert or engage.

Drop-down Navigation or Drop-down Menu. A multilayered style of navigation. There is a top-level of links that is always visible. After hovering the cursor over, clicking on, or tapping on top-level links, a set of subnavigation appears containing more links.

Efficiency. A measurement of how quickly people can complete a desired task. Usually measured in total steps it takes to complete the task and the total time it takes people to complete that task.

Effectiveness. A measurement of how accurately a visitor was able to complete a desired task. This can also be measured based on the quality of the end result of the task the visitor was completing. For instance, did a visitor enter an accurate and complete response into each form field?

Em, Rem, Percent, and Relative Units. There are two means of specifying the height or width of items: absolute units or relative units. On websites, the most common absolute unit is the pixel. Percentage values and em units are relative to the size of a parent item (think of a website's HTML code as a hierarchy where one item owns another). The rem unit is relative to the root (the highest level in the HTML code's hierarchy).

Engagement. An engagement represents all the various ways people interact with and use your website. Technically, a conversion is a type of engagement, but typically conversions are separated out as a special and uniquely important way for people to engage during their visit to your website.

Field. A form field is a means of collecting some kind of information from a visitor. As a few examples, fields can accept text input, take input via checkboxes, or offer a drop-down list of options visitors can select from.

Frontend or Client-Side. Frontend code controls how your website is presented to the visitor in the browser (the browser in this case is the client). This code includes HTML, CSS, or JavaScript. It is called frontend because it is directly in front of people as they visit your website. Also see Backend Code.

Focus Event. A focus event occurs when a visitor moves into a form field. For instance, a visitor might click on a text field to start responding to the question. That click brings the field into focus. Also see Blur Event.

Form. A form is a collection of form fields that let you collect some kind of information from a visitor. Forms are critical for many types of conversions including things like order forms, sign-up forms, sales-inquiry forms, or donation forms.

Heatmap. A heatmap report is a visualization of where visitors are clicking, scrolling, or moving their mouse. Areas with greater activity are highlighted in a different color or are brighter.

Information Foraging. Information foraging is a theory describing how people look for (or hunt for) information. The general idea is people want to exert as little effort as possible finding whatever they are seeking. To help guide people toward the information they are looking for, your website can offer cues to provide hints about what information is contained where.

Label. Labels can be added to a field on a form to clarify what type of information the visitor should put into that field. Labels can also help visitors with visual disabilities better understand each field.

Macro Task. These are the big, obvious tasks people are coming to your website to complete. For example, signing up to use your service, ordering a product, or downloading a key resource would all be macro tasks. Also see Micro Task.

Mental Models. A collections of beliefs and expectations that determine how a person thinks the world (or at least your website) should behave. These beliefs and expectations are based on prior experiences and guesses about the nature of the world.

Micro Task. Micro tasks are smaller, less obvious tasks people come to your website to complete. Those micro tasks might include signing up for your newsletter, clicking on your navigation, or scrolling through a lengthy blog post. Also see Macro Task.

Mistakes and Slips. Both of these are very similar concepts and involve some type of accident occurring. With a mistake, a visitor thought they performed an action correctly but after doing so found out they actually performed that action incorrectly. Slips, however, are purely accidental actions.

Navigation. Navigation represents all the various ways people can move through your website, including the set of navigation links contained at the top of your website, within a sidebar, in the footer, and links within the text of each page.

Navigation Style and Structure. The way in which you choose to organize the various links contained in your website. This might be based on your organization's internal structure (so-called org-chart navigation) or based on different types of visitors (so-called audience-based navigation). Better navigation styles are aligned with the tasks and topics visitors are interested in finding.

Not-Found Error (404 Error). The visitor has requested a file that no longer exists on the website they are visiting. The file might have been removed from the website or maybe the visitor mistyped the address to the page. Technically, when a visitor reached a not-found error, the website server should return a status response code of 404 to indicate the file cannot be found.

Optimization. The process of continually making changes in order to find which of those changes do the best job at achieving the desired outcome—in the case of a website, the outcome is increasing the amount of visitors who engage and convert.

Simplicity. A reduction in complexity in how something is described or presented—what I sincerely hope I have achieved by writing this book.

BIBLIOGRAPHY

Akamai Technologies. "2 Seconds Is Threshold for ECommerce Performance." September 14, 2009. https://www.akamai.com/us/en/about/news/press/2009-press/akamai-reveals-2-seconds-as-the-new-threshold-of-acceptability-for-ecommerce-web-page-response-times.jsp.

Allen, Tim. "Having a CAPTCHA Is Killing Your Conversion Rate." *Moz.* August 5, 2013. https://moz.com/blog/having-a-captcha-is-killing-your-conversion-rate.

Anderson, Chris. *The Long Tail: Why the Future Is Selling Less of More.* Revised and Updated Edition. New York: Hyperion Books, 2008.

Anthony. "How to Make Your Form Error Messages More Reassuring." *UX Movement.* August 23, 2013. http://uxmovement.com/forms/how-to-make-your-form-error-messages-more-reassuring/.

Anthony. "Why Hover Menus Do Users More Harm Than Good." *UX Movement.* March 1, 2011. http://uxmovement.com/navigation/why-hover-menus-do-users-more-harm-than-good/.

Antona, Margherita, Constantine Stephanidis, and International Conference on Human-Computer Interaction (18th; 2016; Toronto, ON), jointly held conference. "Universal Access in Human-Computer Interaction: Methods, Techniques, and Best Practices: 10th International Conference, UAHCI 2016," Held as Part of HCI International 2016, Toronto, ON, Canada, July 17–22, 2016, Proceedings. Part I. n.p., 2016.

Arthur, Charles. "Phone Handsets: The Icon That, Like Floppy Drives, Will Not Die." *The Guardian.* August 23, 2012. https://theguardian.com/technology/blog/2012/aug/23/icons-floppy-phone-handset.

Babich, Nick. "Mobile UX Design: User Errors." June 3, 2016. http://babich.biz/mobile-ux-design-user-errors/.

Bargas-Avila, Javier. "Users Love Simple and Familiar Designs—Why Websites Need to Make a Great First Impression." *Google Research Blog.* August 29, 2012. https://research.googleblog.com/2012/08/users-love-simple-and-familiar-designs.html.

Barry, Nathan. "How to Use Icons to Support Content in Web Design." *Smashing Magazine.* March 3, 2009. http://www.smashingmagazine.com/2009/03/how-to-use-icons-to-support-content-in-web-design/.

Barysevich, Aleh. "How Important Are Tags in 2016 for SEO?" *Search Engine Journal.* February 27, 2016. http://searchenginejournal.com/important-tags-2016-seo/156440/.

Braga, Matthew. "Why Pixel Density Matters More Than Just Screen Size or Resolution." *Tested.com.* June 02, 2010. http://www.tested.com/tech/371-why-pixel-density-matters-more-than-just-screen-size-or-resolution/.

Brinck, Tom, Darren Gergle, Scott D. Wood, David Blythe, and Tom McReynolds. *Usability for the Web: Designing Web Sites That Work.* San Francisco, CA: Morgan Kaufmann, 2001.

Brünken, Roland. *Cognitive Load Theory: Theory and Applications.* Edited by Jan L. Plass, Roxana Moreno, and Roland Brunken. Cambridge: Cambridge University Press, 2010.

Bursztein, Elie, Steven Bethard, Celine Fabry, John C. Mitchell, and Daniel Jurafsky. "How Good Are Humans at Solving CAPTCHAs? A Large Scale Evaluation." In *IEEE Symposium on Security and Privacy*, pp. 399–413, 2010.

Campbell, Ross, Kyle Frost, and Matt Soave. "User Control and Freedom." 2010. http://mattsoave.com/old/cogs187a/iu_site_eval/3control.html.

Cao, Jerry. "The Hidden Power of Inconsistency in Design." *The Next Web.* June 30, 2015. http://thenextweb.com/dd/2015/06/30/the-hidden-power-of-inconsistency-in-design/.

Chen, Mon Chu, John R. Anderson, and Myeong Ho Sohn. "What Can a Mouse Cursor Tell Us More? Correlation of Eye/Mouse Movements on Web Browsing." In *CHI '01 Extended Abstracts on Human factors in computing systems*, pp. 281–82, ACM, 2001.

Cheung, Anna. "8 Proactive Chat Best Practices with Ready-to-Use Scripts-Comm100 Blog." *Comm 100.* February 21, 2016. https://www.comm100.com/blog/proactive-chat-best-practices.html.

Clark, Josh. *Tapworthy: Designing Great IPhone Apps.* Sudbury, MA: O'Reilly Media, Inc., 2010.

Clicktale. "How Website Visitors Scroll and See Your Content." *Clicktale.* December 4, 2007. https://www.clicktale.com/academy/blog/are-your-web-visitors-really-paying-attention/.

Clicktale. "Scrolling Research Report V2.0—Part 1: Visibility and Scroll Reach." *Clicktale.* October 5, 2007. https://www.clicktale.com/academy/blog/clicktale-scrolling-research-report-v20-part-1-visibility-and-scroll-reach/.

Code My Views. "Website Navigation Trends: 10 Tips & Examples." *Code My Views.* September 14, 2015. Last modified July 31, 2016. https://codemyviews.com/blog/website-navigation-trends-10-tips-examples.

Colors On The Web. "Color Contrast." http://www.colorsontheweb.com/Color-Theory/Color-Contrast.

Cousins, Carrie. "Serif vs. Sans Serif Fonts: Is One Really Better Than the Other?" October 28, 2013. https://designshack.net/articles/typography/serif-vs-sans-serif-fonts-is-one-really-better-than-the-other/.

Coyier, Chris. "Responsive Images: If You're Just Changing Resolutions, Use Srcset." *CSS Tricks.* September 30, 2014. https://css-tricks.com/responsive-images-youre-just-changing-resolutions-use-srcset/.

Creative Bloq. *Why Visual Consistency Can Make or Break Your Web Design.* May 22, 2015. http://www.creativebloq.com/web-design/why-visual-consistency-can-make-or-break-your-web-design-51514942.

Cronin, Matt. "Designing Drop-down Menus: Examples and Best Practices." *Smashing Magazine.* March 24, 2009. https://smashingmagazine.com/2009/03/designing-drop-down-menus-examples-and-best-practices/.

Davis, Matthew. "Web Design Basics: Rem vs. Em vs. Px—Sizing Elements in CSS." *Future Hosting.* November 6, 2014. https://www.futurehosting.com/blog/web-design-basics-rem-vs-em-vs-px-sizing-elements-in-css/.

De Valk, Joost. "404 Error Page for WordPress—A Practical Guide." *Yoast.* November 3, 2009. https://yoast.com/dev-blog/404-error-pages-wordpress/.

Eckert, Jay. "Mobile Web Design Patterns—A Look at the Thumb Zone." *Design Trends.* March 14, 2016. http://www.parachutedesign.ca/blog/mobile-design-patterns-a-look-at-the-thumb-zone.

Enders, Jessica. "Well-Designed Error Messages." *Formulate Information Design.* January 11, 2011. http://www.formulate.com.au/blog/well-designed-error-messages.

Facebook for Developers. "The Facebook Crawler." https://developers.facebook.com/docs/sharing/webmasters/crawler.

Fadeyev, Dmitry. "9 Common Usability Mistakes in Web Design." *Smashing Magazine.* February 18, 2009. https://www.smashingmagazine.com/2009/02/9-common-usability-blunders/.

Featherstone, Derek. "Required Form Fields." *Simply Accessible.* October 05, 2005. http://simplyaccessible.com/article/required-form-fields/.

Fisher, Mike. "6 Ways to Improve Confirmation Pages." March 19, 2009. http://completeusability.com/6-ways-to-improve-confirmation-pages/.

Franz, Laura. "Chunking Text with Hierarchy." February 13, 2016. http://typographicwebdesign.com/setting-text/chunking-text-with-hierarchy/.

Fried, Jason. "Ask 37signals: Is It Really the Number of Features That Matter?" October 10, 2007. https://signalvnoise.com/posts/643-ask-37signals-is-it-really-the-number-of-features-that-matter.

Forms that work. "Appearance: Labels, Buttons, Required Fields, and Making Forms Look Good." December 22, 2014. http://www.formsthatwork.com/Appearance.

Foster, James. "Mobile Menu AB Tested: Hamburger Not the Best Choice?" September 27, 2016. http://exisweb.net/mobile-menu-abtest.

Frost, Brad. "Content Parity." March 22, 2012. http://bradfrost.com/blog/mobile/content-parity/.

Fuchs, Thomas. "How to Write a Great Error Message." *Medium.* August 30, 2015. https://medium.com/@thomasfuchs/how-to-write-an-error-message-883718173322.

Gaffney, Gerry. "Why Consistency Is Critical." February 25, 2005. https://www.sitepoint.com/why-consistency-is-critical/.

Gallagher, Danny. "Goldfish Have a Better Attention Span Than You, Smartphone User." May 18, 2015. http://www.cnet.com/news/goldfish-the-actual-fish-not-the-crackers-may-have-a-better-attention-span-than-humans/.

Gardner, Oli. "How to Optimize Contact Forms for Conversions." April 11, 2013. http://unbounce.com/conversion-rate-optimization/how-to-optimize-contact-forms/.

Gocza, Zoltan. "Myth #13: Icons Enhance Usability." UX Myths. http://uxmyths.com/post/715009009/myth-icons-enhance-usability.

Goktürk, Mehmet. "Fitts's Law." *Interaction Design Foundation.* February 03, 2016. https://www.interaction-design.org/literature/book/the-glossary-of-human-computer-interaction/fitts-s-law.

Google. "Create Useful 404 Pages." http://support.google.com/webmasters/answer/93641.

Google. "Googlebot." https://support.google.com/webmasters/answer/182072.

Google. "Image Publishing Guidelines." Accessed November 24, 2016. https://support.google.com/webmasters/answer/114016?hl=en.

Google. "The Importance of Being Seen." November 2014. http://think.storage.googleapis.com/docs/the-importance-of-being-seen_study.pdf.

Google. "The New Multi-screen World: Understanding Cross-platform Consumer Behavior." August 2012. https://ssl.gstatic.com/think/docs/the-new-multi-screen-world-study_research-studies.pdf.

Google Developers. "Mobile Analysis in PageSpeed Insights." April 8, 2015. https://developers.google.com/speed/docs/insights/mobile.

Google Developers. "Mobile SEO Overview." May 16, 2016. https://developers.google.com/webmasters/mobile-sites/mobile-seo/.

Google Developers. "Size Content to Viewport." April 8, 2015. Accessed July 19, 2016. http://developers.google.com/speed/docs/insights/SizeContentToViewport.

Google Developers. "Size Tap Targets Appropriately." April 8, 2015. https://developers.google.com/speed/docs/insights/SizeTapTargetsAppropriately.

Google Developers. "Use Legible Font Sizes." April 8, 2015. https://developers.google.com/speed/docs/insights/UseLegibleFontSizes.

Greenstreet, Karyn. "Should You Put Your Prices on Your Website?" October 06, 2014. http://passionforbusiness.com/blog/should-you-put-your-prices-on-your-website/.

Grieser, Stefanie. "Is Too Much Choice Killing Your Conversion Rates?" April 10, 2014. http://unbounce.com/conversion-rate-optimization/psychology-of-choice-conversion-rates/.

Grigorik, Ilya. "HTTP Caching." *Web Fundamentals* (Google Developers). December 21, 2016. https://developers.google.com/web/fundamentals/performance/optimizing-content-efficiency/http-caching.

Halarewich, Danny. "Single-Page or Multi-Page Checkout: Which Is Better?" *LemonStand*. March 11, 2015. http://blog.lemonstand.com/the-great-debate-single-or-multi-page-e-commerce-checkout/.

Harrison, Chris, Zhiquan Yeo, and Scott E. Hudson. "Faster progress bars: manipulating perceived duration with visual augmentations." In *Proceedings of the SIGCHI conference on human factors in computing systems*, pp. 1545–48. ACM, 2010.

Hornor, Tara. "10 Troublesome Colors to Avoid in Your Advertising." May 8, 2013. https://www.sitepoint.com/10-troublesome-colors-to-avoid-in-your-advertising/.

Hoober, Steven. "How Do Users Really Hold Mobile Devices?" February 18, 2013. http://www.uxmatters.com/mt/archives/2013/02/how-do-users-really-hold-mobile-devices.php.

HTTP Archive. "Trends." Updated November 1, 2016. Accessed November 15, 2016. http://httparchive.org/trends.php?s=All&minlabel=Sep+15+2011&maxlabel=Nov+1+2016.

Huang, Jeff, Ryen White, and Georg Buscher. "User See, User Point: Gaze and Cursor Alignment in Web Search." In *Proceedings of the SIGCHI Conference on Human Factors in Computing Systems*, pp. 1341–50. ACM, 2012.

iA Inc. "The 100% Easy-2-Read Standard." November 17, 2006. https://ia.net/know-how/100e2r.

Impact Information Plain Language Services. "Know Your Readers." *Plain Language at Work Newsletter*. 2013. http://www.impact-information.com/impactinfo/literacy.htm.

International Organization for Standardization (ISO). "Ergonomic Requirements for Office Work with Visual Display Terminals (VDTs)—Part 11: Guidance on Usability." March 19, 1998. Last Modified January 23, 2013. http://www.iso.org/iso/catalogue_detail.htm?csnumber=16883.

Ivanoff, Ada. "Design Minimalism: What, Why & How." June 6, 2014. https://www.sitepoint.com/what-is-minimalism/.

Johnson, Jeff. *Designing with the Mind in Mind: A Simple Guide to Understanding User Interface Design Rules*. Burlington, MA: Morgan Kaufmann Publishers/Elsevier, 2010.

Jones, Russ. "Google Keyword Planner's Dirty Secrets." *Moz*. December 1, 2015. https://moz.com/blog/google-keyword-planner-dirty-secrets.

Jovanovic, Janko. "Web Form Validation: Best Practices and Tutorials." *Smashing Magazine*. July 7, 2009. https://www.smashingmagazine.com/2009/07/web-form-validation-best-practices-and-tutorials/.

Kaushik, Avinash. "Excellent Analytics Tip #13: Measure Macro and Micro Conversions." March 26, 2008. http://www.kaushik.net/avinash/excellent-analytics-tip-13-measure-macro-and-micro-conversions/.

Kaushik, Avinash. "Standard Metrics Revisited: #3: Bounce Rate." August 6, 2007. http://www.kaushik.net/avinash/standard-metrics-revisited-3-bounce-rate/.

Kenny, Tom. "The Use of Logos in Web Design." 2016. http://www.tomkenny.design/articles/the-use-of-logos-in-web-design/.

Kirk, Doug. "Embrace the Controversial: Why You Should Publish Pricing on Your Website." July 25, 2013. http://blog.hubspot.com/marketing/why-publish-pricing-on-website-var.

Kundu, Sourav. "How Does WordPress Caching Work?" *WPExplorer*. October 28, 2013. http://www.wpexplorer.com/wordpress-caching-work/.

Laja, Peep. "PXL: A Better Way to Prioritize Your A/B Tests." *ConversionXL*. September 27, 2016. http://conversionxl.com/better-way-prioritize-ab-tests/.

Laubheimer, Page. "Preventing User Errors: Avoiding Unconscious Slips." *Nielsen Norman Group*. August 23, 2015. https://www.nngroup.com/articles/slips/.

Lebed, Maria. "Best Practice for Proactive Chat. To Engage or Not Engage?" *Provide Support Blog*. October 5, 2013. http://www.providesupport.com/blog/best-practice-for-proactive-chat-to-engage-or-not-engage/.

Liew, Zell. "REM Vs EM—the Great Debate." *Zell Liew*. February 17, 2016. https://zellwk.com/blog/rem-vs-em/.

Lund, Arnold M. "Expert Ratings of Usability Maxims." *Ergonomics in Design: The Quarterly of Human Factors Applications* 5, no. 3 (1997): 15–20.

Maine, Kurt. "List of Tablet and Smartphone Resolutions and Screen Sizes." November 10, 2011. http://binvisions.com/articles/tablet-smartphone-resolutions-screen-size-list/.

Marine, Larry. "Responsive Design Vs. Task-Oriented UX Design." *Search Engine Watch.* February 27, 2014. https://searchenginewatch.com/sew/how-to/2331203/responsive-design-vs-task-oriented-ux-design.

MarketingExperiments. "Friction: Are Your Webpages Rubbing Customers the Wrong Way?" *Marketing Experiments.* July 2, 2007. http://www.marketingexperiments.com/blog/research-topics/site-design/friction-are-your-webpages-rubbing-customers-the-wrong-way.html.

Markman, Art. "How Distraction Can Disrupt You." *Psychology Today.* March 18, 2014. https://www.psychologytoday.com/blog/ulterior-motives/201403/how-distraction-can-disrupt-you.

Massey, Brian. "7 Best Practices for Your Exit-Intent Popovers and Popups." *Conversion Sciences.* January 20, 2015. http://conversionsciences.com/blog/7-best-practices-using-exit-intent-popovers/.

Mathew, George. "Opt-in Pop-Ups: Are They Any Good?" *The Daily Egg.* August 18, 2014. https://blog.crazyegg.com/2014/08/18/opt-pop-ups/.

McGovern, Gerry. "Search Articles." *New Thinking.* January 29, 2012. http://www.gerrymcgovern.com/new-thinking/why-audience-navigation-usually-doesn%E2%80%99t-work.

McGrane, Karen. "Your Content, Now Mobile." *A List Apart.* November 5, 2012. http://alistapart.com/article/your-content-now-mobile.

McLeod, Saul. "Cognitive Dissonance." 2014. http://www.simplypsychology.org/cognitive-dissonance.html.

Meunier, Bryson. "82% of Sites Use Responsive Web Design in 2015? Try 11.8%." January 15, 2015. http://marketingland.com/82-sites-use-responsive-web-design-2015-try-11-8-114050.

Miksen, Chris. "What Is the Difference Between Efficiency and Effectiveness in Business?" *Chron* (Chron.com). December 2011. http://smallbusiness.chron.com/difference-between-efficiency-effectiveness-business-26009.html.

Miller, Brad. "404 Page Best Practices." *Search Engine Watch.* September 11, 2013. https://searchenginewatch.com/sew/how-to/2293339/404-page-best-practices.

Miller, Jeremy. "Should You Publish Pricing on Your Website?" June 18, 2015. https://stickybranding.com/should-you-publish-pricing-on-your-website/.

Mind Tools Editorial Team. "Cognitive Load Theory: Making Learning More Effective." April 9, 2016. https://www.mindtools.com/pages/article/cognitive-load-theory.htm.

National Eye Institute. "Facts about Color Blindness." February 2015. https://nei.nih.gov/health/color_blindness/facts_about.

New York Department of Labor. "User Control Freedom—User Experience Guide." Accessed November 18, 2016. http://www.labor.ny.gov/ux/principles-user-control-freedom.html.

Nielsen, Jakob. "10 Best Application UIs." *Nielsen Norman Group.* August 12, 2008. https://www.nngroup.com/articles/10-best-application-uis/.

Nielsen, Jakob. "10 Heuristics for User Interface Design." *Nielsen Norman Group*. 1998. https://www.nngroup.com/articles/ten-usability-heuristics/.

Nielsen, Jakob. "Drop-down Menus: Use Sparingly." *Nielsen Norman Group*. November 12, 2000. https://www.nngroup.com/articles/drop-down-menus-use-sparingly/.

Nielsen, Jakob. "Error Message Guidelines." *Nielsen Norman Group*. June 24, 2001. https://www.nngroup.com/articles/error-message-guidelines/.

Nielsen, Jakob. "Horizontal Attention Leans Left." *Nielsen Norman Group*. April 06, 2010. https://www.nngroup.com/articles/horizontal-attention-leans-left/.

Nielsen, Jakob. "Information Foraging: Why Google Makes People Leave Your Site Faster." *Nielsen Norman Group*. June 30, 2003. http://nngroup.com/articles/information-scent/.

Nielsen, Jakob. "Mental Models and User Experience Design." October 18, 2010. https://www.nngroup.com/articles/mental-models/.

Nielsen, Jakob. "Mobile Content: If in Doubt, Leave It Out." *Nielsen Norman Group*. October 10, 2011. http://www.nngroup.com/articles/condense-mobile-content/.

Nielsen, Jakob. "Mouse vs. Fingers as Input Device." *Nielsen Norman Group*. April 10, 2012. http://www.nngroup.com/articles/mouse-vs-fingers-input-device/.

Nielsen, Jakob. "Severity Ratings for Usability Problems: Article by Jakob Nielsen." *Nielsen Norman Group*. January 1, 1995. https://www.nngroup.com/articles/how-to-rate-the-severity-of-usability-problems/.

Niggulis, Ott. "In Defense of the Email Popup." *ConversionXL*. June 26, 2014. http://conversionxl.com/popup-defense/.

Niggulis, Ott. "Which Color Converts the Best?" *ConversionXL*. January 16, 2013. http://conversionxl.com/which-color-converts-the-best/.

Norman, Don A. *The Design of Everyday Things*. Revised and Expanded edition. New York: Basic Books, 2013.

Occupational Safety and Health Administration (OSHA). "Workstation Components Monitors." February 4, 2015. https://osha.gov/SLTC/etools/computerworkstations/components_monitors.html.

Ohye, Maile. "Site Performance For Webmasters." *Google Webmasters*. YouTube video, 12:10. Posted May 2010. https://www.youtube.com/watch?v=OpMfx_Zie2g.

Osborne, Tom. "Color Contrast for Better Readability." *Viget*. March 3, 2015. https://www.viget.com/articles/color-contrast.

Pamental, Jason. "A More Modern Scale for Web Typography." January 15, 2015. http://typecast.com/blog/a-more-modern-scale-for-web-typography.

Page, Rich. "A/B, Split and Multivariate Test Duration Calculator." *Visual Website Optimizer*. July 13, 2016. http://rich-page.com/website-optimization/how-to-test-and-improve-your-website-if-your-traffic-is-too-low-for-ab-testing/.

Patel, Neil. "9 Landing Page Elements That Need to Die." *The Daily Egg*. July 15, 2014. https://blog.crazyegg.com/2014/07/15/remove-landing-page-elements/.

Pernice, Kara and Raluca Budiu. "Hamburger Menus and Hidden Navigation Hurt UX Metrics." *Nielsen Norman Group.* June 26, 2016. https://www.nngroup.com/articles/hamburger-menus/.

Phan, Doantam. "Helping Users Easily Access Content on Mobile." Google Webmaster Central Blog. August 23, 2016. https://webmasters.googleblog.com/2016/08/helping-users-easily-access-content-on.html.

Phillips, Patrick. "The Entire World Hates CAPTCHA. So Why's It Still Here?" *Patrick's Place.* July 4, 2011. http://www.patrickkphillips.com/blogging/entire-world-hates-captcha-whys-still/.

Piejko, Pawel. "Most Popular Smartphone Screen Resolutions in 2015." July 2, 2015. https://deviceatlas.com/blog/most-popular-smartphone-screen-resolutions-2015.

Pirolli, Peter and Stuart Card. "Information Foraging." *Psychological Review* 106, no. 4 (1999): 643–75. doi:10.1037//0033-295x.106.4.643.

Preibusch, Sören, Kat Krol, and Alastair R. Beresford. "The Privacy Economics of Voluntary Over-disclosure in Web Forms." In *The Economics of Information Security and Privacy*, edited by Rainer Böhme, 183–209. Berlin, Heidelberg: Springer, 2013.

Qualaroo. "Reducing Bounce and Exit Rates." 2016. https://qualaroo.com/beginners-guide-to-cro/reducing-bounce-and-exit-rates/.

Rabida, Kevin Mark. "The Difference Between Minimalism and Simplicity." February 23, 2015. http://www.ucreative.com/articles/minimalism-simplicity-difference/.

RapidTables. "Screen Resolution Statistics." February 2014. http://www.rapidtables.com/web/dev/screen-resolution-statistics.htm.

Rettner, Rachael. "How Your Brain Works on Autopilot." *Live Science.* June 9, 2010. http://www.livescience.com/6557-brain-works-autopilot.html.

Richman, Danny. "How Online Visitors Feel about Pop-Ups." July 20, 2016. https://www.seotraininglondon.org/should-use-popups-website/.

Rosenfield, Mark. "Computer Vision Syndrome: A Review of Ocular Causes and Potential Treatments." *Ophthalmic and Physiological Optics* 31, no. 5 (2011): 502–15.

Sauro, Jeff. "Rating the Severity of Usability Problems: MeasuringU." *MeasuringU.* July 30, 2013. http://www.measuringu.com/blog/rating-severity.php.

Schade, Amy. "Don't Force Users to Register Before They Can Buy." *Nielsen Norman Group.* July 15, 2015. https://www.nngroup.com/articles/optional-registration/.

Schade, Amy. "The Fold Manifesto: Why the Page Fold Still Matters." *Nielsen Norman Group.* February 1, 2015. https://www.nngroup.com/articles/page-fold-manifesto/.

Schwartz, Barry. "Is the Famous 'Paradox of Choice' a Myth?" PBS (PBS NewsHour). February 20, 2014. http://www.pbs.org/newshour/making-sense/is-the-famous-paradox-of-choic/.

Sexton, Patrick. "Use Legible Font Sizes." March 5, 2016. https://varvy.com/mobile/legible-font-size.html.

Sharma, Himanshu. "MSc Digital Marketing Online." November 6, 2012. http://www.webanalyticsworld.net/2012/11/how-to-separate-user-engagement-from-distraction.html.

Sherwin, Katie. "The Magnifying-Glass Icon in Search Design: Pros and Cons." *Nielsen Norman Group*. February 23, 2014. https://www.nngroup.com/articles/magnifying-glass-icon/.

Shet, Vinay. "Are You a Robot? Introducing 'No CAPTCHA reCAPTCHA.'" *Google Security Blog*. December 3, 2014. https://security.googleblog.com/2014/12/are-you-robot-introducing-no-captcha.html.

Shneiderman, Ben. "The Eight Golden Rules of Interface Design." 2010. https://www.cs.umd.edu/users/ben/goldenrules.html.

Silverman, Elisa. "Do the New Captchas Affect Conversion Rate?" *Pagewiz*. February 10, 2016. http://www.pagewiz.com/blog/conversion-rate-optimization/new-captchas-conversion-rate.

Six, Janet M. "Designing for Senior Citizens | Organizing Your Work Schedule." May 17, 2010. http://www.uxmatters.com/mt/archives/2010/05/designing-for-senior-citizens-organizing-your-work-schedule.php.

Smith, Jared. "WCAG 2.0 and Link Colors." *WebAIM*. July 24, 2009. http://webaim.org/blog/wcag-2-0-and-link-colors/.

Smith, Jeremy. "How to Win the War Against Conversion Friction." *Kissmetrics blog*. June 10, 2014. https://blog.kissmetrics.com/war-against-conversion-friction/.

Smith, Jeremy. "Optimizing For Conversions: Increasing Relevance." *Jeremy Said*. April 7, 2016. https://jeremysaid.com/blog/conversion-triangle-series-increasing-relevance/.

Souders, Steve. "The Performance Golden Rule." *SteveSouders.com*. February 10, 2012. https://www.stevesouders.com/blog/2012/02/10/the-performance-golden-rule/.

StatCounter. "Global Stats." http://gs.statcounter.com/#resolution-ww-monthly-201512-201612. Screen Resolution graph, all device types, 2016 annual data.

Strizver, Ilene. "Serif vs. Sans for Text in Print." April 2013. https://www.fonts.com/content/learning/fontology/level-1/type-anatomy/serif-vs-sans-for-text-in-print.

Sweller, John, Paul Ayres, and Slava Kalyuga. *Cognitive Load Theory*. New York: Springer, 2011.

Tackett, John. "Landing Page Optimization: What a 29% Drop in Conversion Can Teach You about Friction." *Marketing Experiments*. July 7, 2014. http://www.marketingexperiments.com/blog/research-topics/landing-page-optimization-research-topics/drop-in-conversion-teach-friction.html.

Tennant, D. Bnonn. "16 Pixels: For Body Copy. Anything Less Is A Costly Mistake." *Smashing Magazine*. October 7, 2011. https://smashingmagazine.com/2011/10/16-pixels-body-copy-anything-less-costly-mistake/.

Thercaux, Olivier and Susan Lesch. "Care with Font Size." April 9, 2010. http://.w3.org/QA/Tips/font-size.

Trepess, David. "Information Foraging Theory." Accessed July 9, 2016.
http://interaction-design.org/literature/book/the-glossary-of-human-computer-interaction/information-foraging-theory.

Unbounce. "Call to Action Design | Landing Page Conversion | Part 3." 2016.
http://thelandingpagecourse.com/call-to-action-design-cta-buttons/.

Urban, Diana. "Should You Remove Navigation from Your Landing Pages? Data Reveals the Answer." January 9, 2014.
http://blog.hubspot.com/marketing/landing-page-navigation-ht.

Usability First. "Usability Glossary—Graceful Flow." May 11, 2010.
http://www.usabilityfirst.com/glossary/graceful-flow/.

US General Services Administration (GSA). "Quick Reference Guide to Section 508 Requirements and Standards." Accessed July 22, 2016.
https://section508.gov/content/quick-reference-guide.

W3C. "Contrast (Minimum) Understanding Success Criterion 1.4.3."
Understanding WCAG 2.0. November 10, 2016.
https://www.w3.org/TR/UNDERSTANDING-WCAG20/visual-audio-contrast-contrast.html.

Wallace, Tracey. "Why Your Ecommerce Conversion Rate Determines Your Site's Success." *Big Commerce.* April 1, 2016.
https://www.bigcommerce.com/blog/conversion-rate-optimization/.

Wargo, Eric. "How Many Seconds to a First Impression?" July 1, 2006.
https://psychologicalscience.org/observer/how-many-seconds-to-a-first-impression.

WebAIM. "WebAIM: Color Contrast Checker."
http://webaim.org/resources/contrastchecker/.

Weinschenk, Susan. "The Secret to Designing an Intuitive UX." *UX Magazine.*
October 8, 2011. https://uxmag.com/articles/the-secret-to-designing-an-intuitive-user-experience.

Wendelin, Eric. "Change Text Size On Click With JavaScript." *David Walsh Blog.*
January 31, 2008. https://davidwalsh.name/change-text-size-onclick-with-javascript.

Wesley, Daniel. "How the World Spends Its Time Online." June 16, 2010.
https://www.creditloan.com/blog/how-the-world-spends-its-time-online/.

West, Matt. "An Introduction to Perceived Performance - Treehouse Blog." April 23, 2014. http://blog.teamtreehouse.com/perceived-performance.

Whitenton, Kathryn. "Website Logo Placement for Maximum Brand Recall."
Nielsen Norman Group. February 21, 2016. https://nngroup.com/articles/logo-placement-brand-recall/.

Work, Sean. "How Loading Time Affects Your Bottom Line." *Kissmetrics blog.*
April 30, 2011. https://blog.kissmetrics.com/loading-time/.

Wroblewski, Luke. "Inline Validation in Web Forms." *A List Apart.* September 1, 2009. http://alistapart.com/article/inline-validation-in-web-forms.

Wroblewski, Luke. "Responsive Navigation: Optimizing for Touch Across Devices."
LukeW. November 2, 2012. http://www.lukew.com/ff/entry.asp?1649.

Yankee, Everyl. "Object-Focused vs. Task-Focused Design." *UX Mastery.*
December 7, 2012. http://uxmastery.com/object-focused-vs-task-focused/.

ABOUT THE AUTHOR

Matthew Edgar is web consultant specializing in analytics, technical marketing, and conversion optimization. Since 2001, he has helped hundreds of businesses, startups, and nonprofits grow through a process of analyzing and improving their websites and online presence. Matthew regularly speaks at conferences and teaches workshops about data, analytics, and technical marketing subjects. He is a partner and consultant at Elementive (Elementive.com) and blogs at MatthewEdgar.net.

www.ingramcontent.com/pod-product-compliance
Lightning Source LLC
Chambersburg PA
CBHW071544200326
41519CB00021BB/6609